TAKING THE LEAP

TAKING THE LEAP

The Methods and Tools of the Linked Engineering and Manufacturing Platform (LEAP)

Edited by

DIMITRIS KIRITSIS
EPFL, Lausanne, Switzerland

ELSEVIER

Amsterdam • Boston • Heidelberg • London
New York • Oxford • Paris • San Diego
San Francisco • Singapore • Sydney • Tokyo

Academic Press is an imprint of Elsevier

Academic Press is an imprint of Elsevier
125 London Wall, London EC2Y 5AS, United Kingdom
525 B Street, Suite 1800, San Diego, CA 92101-4495, United States
50 Hampshire Street, 5th Floor, Cambridge, MA 02139, United States
The Boulevard, Langford Lane, Kidlington, Oxford OX5 1GB, UK

British Library Cataloguing-in-Publication Data
A catalogue record for this book is available from the British Library

Library of Congress Cataloging-in-Publication Data
A catalog record for this book is available from the Library of Congress

ISBN: 978-0-12-805263-1

For information on all Academic Press publications
visit our website at https://www.elsevier.com/

Publisher: Joe Hayton
Acquisition Editor: Brian Guerin
Editorial Project Manager: Carrie Bolger
Production Project Manager: Julie-Ann Stansfield
Designer: Matthew Limbert

CONTENTS

CONTRIBUTORS

D. Ahlers
NTNU, Trondheim, Norway

A. Buda
Aalto University, Espoo, Finland

J. Cassina
HOLONIX S.R.L., Meda, Italy

D. Cerri
Polytechnic of Milan, Milan, Italy

M. Cocco
Polytechnic of Milan, Milan, Italy

E. Coscia
HOLONIX S.R.L., Meda, Italy

S. El Kadiri
EPFL, Lausanne, Switzerland

K. Främling
Aalto University, Espoo, Finland

G. Iversen
Aker Solutions AS, Norway

D. Kiritsis
EPFL, Lausanne, Switzerland

K. Kristensen
NTNU, Trondheim, Norway

J. Krogstie
NTNU, Trondheim, Norway

S. Kubler
Aalto University, Espoo, Finland

M. Mehrpoor
NTNU, Trondheim, Norway

A. Milicic
EPFL, Lausanne, Switzerland

A. Mocan
SAP SE, Dresden, Germany

K. Pardalis
UBITECH LTD, Athens, Greece

S. Parrotta
HOLONIX S.R.L., Meda, Italy

F. Perales
Trimek S.A., Spain

E. Peukert
University of Leipzig, Leipzig, Germany

M. Rossi
Polytechnic of Milan, Milan, Italy

S. Terzi
Polytechnic of Milan, Milan, Italy

C. Wartner
University of Leipzig, Leipzig, Germany

PREFACE

LINKED ENGINEERING AND MANUFACTURING: A KEY CHALLENGE FOR INDUSTRY 4.0

Manufacturing is the driving force of Europe's economy, contributing over €6,553 billion in GDP and providing more than 30 million jobs. Although a strong manufacturing sector is vital to European economic growth and stability, this sector is facing increasingly difficult challenges. The economic crisis has decreased industry output by around 20% while global competition is growing dramatically. Furthermore, new trends and paradigms such as an increasing demand for sustainable manufacturing and mass customization are increasing. ICT is the key enabler for coping with these changes to promote engineering and manufacturing excellence as a driver for European success.

The current ICT for the manufacturing landscape is characterized by scattered data formats, tools, and processes dedicated to different phases in the product lifecycle: in the concept phase of a product common tools, such as MS PowerPoint are used, while later on more specialized tools are used, such as CAx solutions, PLM and ERP systems, and so on. Moreover, the flow of information is closely aligned with the product lifecycle (ie, information from the design phase goes into the manufacturing phase, and can return in the opposite direction in the form of user feedback into designs that have broken or been neglected). Due to the diversity of tools and data formats, manufacturing struggles to cope with new trends in this area. For example, both the trend to mass customization and the demand for increased sustainability require a tight integration of the design, manufacturing, and usage phases of a product, which is currently not in place. The rise of Web 2.0 leads to precious information, manifested in Web 2.0 channels such as blogs and forums, being created directly by prospective or existing users of a given product. But this sort of information is far from having any impact to the design or manufacturing phase of a product.

In summary, what is clearly missing in the current ICT landscape for manufacturing is an integrated, holistic view on data, persons, and processes across the full product lifecycle. As experiences of the past show, a *tight* integration of all tools used throughout a product lifetime is not feasible. For this reason, the EU funded FP7 FoF project 284613 LinkedDesign has developed a Linked Engineering and mAnufacturing Platform (LEAP) for manufacturing to address the current shortcomings.

The LEAP has been designed as an integrated information system for manufacturing design that federates all of the relevant product lifecycle information, independent of its format, location, originator, and time of creation, with the objective to boost today's engineers by providing an integrated, holistic view on data, persons, and processes across the full product lifecycle as a vital resource for the outstanding competitive design of novel products and manufacturing processes.

The book "Taking the LEAP" provides a complete and detailed view of the main results of the LinkedDesign project, which have been integrated in the LEAP.

OBJECTIVES AND OUTLINE OF THE BOOK

The target of this book is to present the results of the LinkedDesign project to the Industry 4.0 community. The content is based on the main technical deliverables of the project and organized as follows:

Chapter 1 "Introduction" by D. Kiritsis and A. Mocan presents the main concepts of the LinkedDesign approach and the main elements of the LEAP.

Chapter 2 "LEAP Data and Knowledge Integration Infrastructure" by E. Peukert and C. Wartner, introduces the Data and Knowledge Integration infrastructure of LEAP.

Chapter 3 "LEAP Semantics" by S. El Kadiri, A. Milicic, K. Pardalis, and E. Peukert proposes the LEAP semantic model that is built on the basis of the design of an upper ontology describing the LEAP domain and its specialization to the three industrial use-cases: Volkswagen, COMAU, and Aker Solutions.

Chapter 4 "LEAP Product and Manufacturing Design Support System" by D. Cerri and S. Terzi presents the development of the so called "LEAP model for design support" and the Life Cycle Costing (LCC) and Life Cycle Assessment (LCA) methodologies as they have been used in the COMAU case.

Chapter 5 "LEAP Collaboration System" by K. Kristensen, J. Krogstie, D. Ahlers, and M. Mehrpoor, presents the models, concepts, elements, and technology components that—when combined and structured in a meaningful way to teams of end users—enable companies to execute split location engineering projects in a way that represents a competitive advantage.

Chapter 6 "LEAP Interoperability Standards" by K. Främling, A. Buda, S. Kubler, J. Cassina, E. Coscia, S. Parrotta, S. Terzi, and D. Cerri, presents interoperability standards developed for this purpose, which are published by The Open Group: Open Messaging Interface (O-MI), Open Data Format (O-DF) and describes the design principles and provides a description of the standards, including implementation principles and examples of real-life implementations.

Chapter 7 "LEAP Virtual Obeya" by M. Rossi, M. Cocco, S. Terzi, K. Kristensen, S. Parrotta, J. Krogstie, and D. Ahlers presents the LEAP Virtual Obeya, a flexible concept that incorporates lean thinking and enables new visual project management approaches in teams engaged in specific processes, such as innovation and engineering design, offering enhanced support for specific tasks such as problem solving, decision making, coediting, and issue/task management.

Finally, Chapter 8 "LEAP Use Cases" by A. Milicic, S. El Kadiri, F. Perales, S. Parrotta, and G. Iversen describes the LinkedDesign use cases and explains how the LEAP platform is being exploited for given tasks.

The book's editor is especially grateful to all of the contributors who have invested their time to produce a high-quality work, introducing an original contribution in the area of ICT enabling technologies for manufacturing. This high-level panel of experts gives this collective book a unique coverage of a content which is crucial for the successful implementation of the Industry 4.0 paradigm, and which could not have been achieved without their active involvement. Last but not the least, the editor, on behalf of all contributors of this book, is grateful to DG CNECT of the European Commission for providing the funding to the LinkedDesign consortium under the FP7 program.

Dimitris Kiritsis
Editor

CHAPTER 1

Introduction

D. Kiritsis* and A. Mocan**
*EPFL, Lausanne, Switzerland
**SAP SE, Dresden, Germany

Contents

Without doubt, manufacturing remains vitally important for the economy of the European Union. According to Ref. [1], before the latest economic crisis, manufacturing contributed some 17.1% of GDP and accounted for some 22 million jobs (2007). By taking the sectors that are directly related to manufacturing into account (eg, transport), [1] this figure rises to 47% of GDP. However, the influence of the recent economic crisis on the European manufacturing industry has decreased output by around 20%, while global competition has grown dramatically. This has led to increasing pressure on the industry. Moreover, several new trends and paradigms, such as sustainable manufacturing and mass customization, have started to emerge. Consequently, the manufacturing industry is facing significant structural changes.

The key enabler for coping with these changes will be ICT, due to its strong impact on innovation and productivity [2–4]. Currently, ICT in the manufacturing industry is characterized by scattered data formats, and tools and processes that are dedicated to different phases in the product lifecycle. For example, common tools such as MS PowerPoint are often used in the concept phase of a product, while specialized CAx solutions, such as PLM and ERP systems, and so on, are used later on. Moreover, the flow of information is closely aligned to the product lifecycle (ie, information from the design phase goes into the manufacturing phase, while it also returns in the opposite direction in the form of user feedback from designs that are broken

Taking the LEAP
http://dx.doi.org/10.1016/B978-0-12-805263-1.00001-3

or neglected). Due to the wide diversity of tools and data formats that are available, manufacturing has struggled to cope with new trends in this area. For example, the trend to mass customization and the demand for increased sustainability require a tight integration of the design, manufacturing, and usage phases of a product, which is currently not in place. The rise of Web 2.0 has led to the growth of information that is created directly by prospective or existing users of a given product, which is manifested in Web 2.0 channels such as blogs and forums. Despite its importance, this sort of information is far from having an impact on the design or manufacturing phase of a product.

To summarize, what is clearly missing in the current ICT landscape for manufacturing is an integrated, holistic view of data, persons, and processes across the full product lifecycle. As previous experience shows, a *tight* integration of all tools used throughout a product lifetime is not feasible. For this reason, LinkedDesign has developed a Linked Engineering and mAnufacturing Platform (LEAP) that the manufacturing industry can use to address the current shortcomings. The aims of LinkedDesign can be briefly described as follows:

Data federation: LEAP federates all of the relevant information across trusted sources in the product lifecycle, independent of its format, location, and origination time.

1. *Context-driven access and analysis of federated information*: Besides unified access to the integrated information, LEAP also provides specific means to analyze the integrated information.
2. *User collaboration*: LEAP is user centric rather than information centric. To foster collaboration between users across different disciplines, LEAP will use and extend lean engineering principles and implement a collaboration workbench that will enable effective internal and external collaboration.

Feedback into existing systems: In addition to pulling data from existing data sources and systems, LEAP will provide tight connections to the federated systems (eg, CAx) so that it can push enriched information back to them.

In the following sections, we elaborate on these four capabilities of the envisioned engineering platform.

1.1 OBJECTIVE 1: DATA FEDERATION

The challenge: Relevant information about a product is scattered in different formats and locations, and searching for information across different tools and data formats is a tedious and error-prone process. Finally, more and more information about a product is currently not being stored in one of the specific structured document formats (eg, relational data, XML data, Excel files, CAx files, etc.) of specific tools, but is hidden in unstructured formats, such as (the text of) office documents, emails, or (enterprise-internal) forums and blogs. In fact, the ratio of unstructured data amongst all data is estimated to be 80–85%, leaving structured data far behind in second place. To summarize: Information resides in a large number of disconnected information silos, and the user might not have genuine access or—even worse—might not be aware of these silos. Thus, a single point of entry for all relevant product information—independent of its format, location, or originator and the phase of the product lifecycle in which the information is created—is clearly missing in the current IT environment. Moreover, implicit (new) information which could be derived from existing information by embracing its mutual relationships and interdependencies remains invisible (this is, important in the recent Volkswagen use case).

The LinkedDesign approach: On its lowest technical level, LinkedDesign serves as a platform which federates all of the relevant product data. This layer will embrace industrial-driven protocols, such as open data [5]. A domain ontology serves as a unified schema to (technically) access all of the federated data; at the same time, it is flexible enough for schema extensions to model domain specifics. Schema matching algorithms are used to map different data sources to the domain ontology, thus making the underlying source schema transparent to the higher layers and the end users. Object matching algorithms are used for duplicate detection and data fusion. Finally, with lightweight reasoning facilities, implicit knowledge is made explicit from the federated data.

1.2 OBJECTIVE 2: CONTEXT-DRIVEN ACCESS AND THE ANALYSIS OF FEDERATED INFORMATION

The challenge: In the previous section we have argued that engineers and manufacturers currently do not have access to all of the product information that they need. Nevertheless, engineers are already facing a quite contrary problem, namely an overload of unrequired, futile, or irrelevant information. According to Ref. [6], it is estimated that information workers spend 13 h a week on information gathering and analysis. In addition, according to Ref. [7], IDC estimates that an enterprise employing 1000 knowledge workers wastes at least $2.5–$3.5 million per year searching for nonexistent information, failing to find existing information, or recreating information that cannot be found. In fact, a system that federates all of the product information sources and which provides a single point of entry might resolve the information overload problem; however, due to the sheer amount of available information, it would not be able to do this by itself. Even if such a system provides a perfect search engine, which searches across all of the relevant information silos, it may not be sufficient to provide the specific information that the user is looking for Ref. [8]. First, search engines usually neglect the context and the role of the user, and thus do not address his or her particular needs. Second, search engines only support ★direct★ search for specific information, they do not allow us to navigate through the information space in an explorative manner. Finally, data analysis tailored to the manufacturing domain (eg, analysis of product designs and quality control results to derive optimal manufacturing configurations) is also not supported.

The LinkedDesign approach: LinkedDesign addresses these needs as follows. First, the information to be displayed to the user will be filtered in advance based on the role of the user and the context of the information access. Second, semantic technologies are used to support different means of information access, both for direct search (eg, keyword search and semantic search) and to explore the information space (eg, faceted search or graph-based information representations), which can reduce the time spent on knowledge acquisition by up to 50%. Finally, the front-end of LEAP has been developed as a Virtual Obeya, which is a virtual control center where all of the information is integrated and shared at desks for each involved actor, it also provides user-centered or role-based workspaces.

1.3 OBJECTIVE 3: USER COLLABORATION

The challenge: With the previous two objectives, it can be seen that the LinkedDesign platform is clearly an improvement of existing solutions for data federation (eg, enterprise search systems or MS SharePoint), but they still only focus on an integrated access to the data and they do not attempt to provide an integration of people or processes. However, networking and knowledge exchange between people are becoming increasingly important. As already outlined, this is particularly important to improve manufacturing processes. A closer integration of the different phases in a product lifecycle assumes the collaboration of the respective actors of these phases (eg, designers, engineers, manufacturers, etc.), but the current collaboration platforms and approaches are still fragmented and disconnected, and they do not support true lean engineering to the desired extent. In the last decade, a variety of social network tools have been developed, both in the web sphere (eg, social networking platforms) and in enterprise internal collaboration platforms (Enterprise 2.0 tools and platforms like semantic media wikis used within companies). However, while these tools represent potentially very valuable, rich sources of information, none of these tools or platforms is currently well integrated in lean engineering processes. As a result, they fail to effectively fill the knowledge gaps that are observable in current core engineering systems, such as handling mostly structured data and information.

The LinkedDesign approach: LinkedDesign attempts to combine the best practices and features from information integration systems and collaboration platforms through a new approach that extends lean collaborative engineering with a collaboration workbench and rich, intraorganizational social network functionalities. The goal is to find competences and foster know-how exchange inside a company, especially across departments and domain responsibilities, which will reduce the number of incidents where experiences are not transferred to subsequent projects by approximately 50%.

1.4 OBJECTIVE 4: FEEDBACK INTO EXISTING SYSTEMS

The challenge: LinkedDesign does not simply aggregate information from different sources and present the aggregated role- and context-based information to the user. Due to the envisioned features of LinkedDesign as outlined in the last objective (ie, embracing relationships between different information pieces, applying reasoning facilities in order to derive new information, adding information based on tagging and collaboration support), the information presented to the user is more than the sum of the aggregated source information. For this reason, LinkedDesign not only pulls information from various sources but it also integrates push mechanisms and other bidirectional integration approaches to the source system.

The LinkedDesign approach: In LinkedDesign, we focus on the early phases of the product lifecycle and develop a tight and seamless integration with CAx-Systems. By doing so, CAx systems are able to become knowledge-based engineering supported applications. A rule interchange format was developed which enables an application-independent management of rules and constraints on product design. In this respect, both human users as well as existing systems could benefit from LinkedDesign. By abolishing the laborious and often error-prone manual integration of knowledge in engineering systems, the overall development time can be reduced tremendously, while the number of automatically supported systems will be multiplied by a factor of at least 5.

1.5 THE LINKEDDESIGN APPROACH

The central goal of LinkedDesign has been to develop LEAP that enables the context-driven and collaborative access to data, information, and knowledge from both the engineering and the manufacturing world. As pointed out in Objective 1, relevant information are scattered in different formats and locations. By using LEAP, engineers are able to access and integrate (disconnected) information sources in a convenient way. While the engineering phase is mainly documented in unstructured information sources, the data management within manufacturing processes is rather structured (Table 1.1).

To enable the seamless integration of both aspects, the *Data Integration* of LEAP applies novel web-scale object matching algorithms to deduplicate multiple logic representations of single real-world objects. For example, Data Integration can find and link references to one specific product distributed over a PLM database, including the CAD-file that defines the structural product design, the marketing presentation on that product, the product-announcement on the enterprise web page, and customer ratings in a web forum. Furthermore, object and schema matching strategies are used to identify the relationships between distinct objects as input for the creation of the *LinkedDesign Knowledge Network* (or the so-called Smart-Link repository). While the LinkedDesign ontology defines the unified domain schema of this knowledge network, schema matching algorithms are used to automatically map and integrate local data sources to this (global)

Table 1.1 Examples of structured and unstructured information sources

Unstructured information sources	Structured information sources
• Design illustrations and images • Technical component descriptions • Material specifications and reports • Emails and other communication logs • Blogs, product reviews, news	• ERP, CRM, PLM, MES systems, etc. • Business process models • (Public) data services • Sensor data, KPIs • Customer ratings

schema, so that the LinkedDesign Knowledge Network will serve as a single access point for all LEAP components and users.

While developments in the fields of data integration and knowledge network are domain-independent and can be applied to arbitrary application scenarios, the further LEAP components present manufacturing and engineering specific functionality. *Knowledge Exploitation* has been developed as a flexible data analysis framework where domain and/or application specific analytic bundles can be deployed. LinkedDesign provides decision support bundles for the optimal design of manufacturing processes and products, as well as a bundle which enables the sentiment analysis to mine customer opinions in relation to product specifications and design decisions.

The *Automated Design* module extends the current approach of knowledge-based engineering to provide a seamless system and workflow integration (see Objective 4). First, it enables the automated formulation and extraction of parameters, rules, and constraints on optimal products and process design from the LinkedDesign Knowledge Network. Second, the developed rule interchange format serves as an interface specification to transfer the extracted design knowledge into existing engineering and manufacturing systems and workflows. For example, based on manufacturing experiences in welding crosspieces, the engineer's CAD tool is able to present optimal angle designs that provide good stability.

The context-driven data access allows LEAP to identify the users' working contexts to provide important data relevant in the respective working situations, as illustrated in Fig. 1.1.

The LinkedDesign solution provides an infrastructure to improve the quality of engineering decisions, to design manufacturing systems, and to evaluate the quality of the resulting products and production processes. It leverages the combined power of semantics and open innovation approaches to improve quality and reduce development lead times. Finally, LEAP is applicable across a wide range of industrial sectors, and is designed to support the entire lifecycle of a product, including product-related services and its final user experience. Hence, LinkedDesign involves several actors and different systems, which are shown in Fig. 1.2. Actually, the LinkedDesign engineering platform involves three tiers of users, as described below.

First-tier users: These users will actively design and engineer the product and the related processes, defining the following:

- The specifications and market needs (marketing);
- The design and the artistic form of the product;

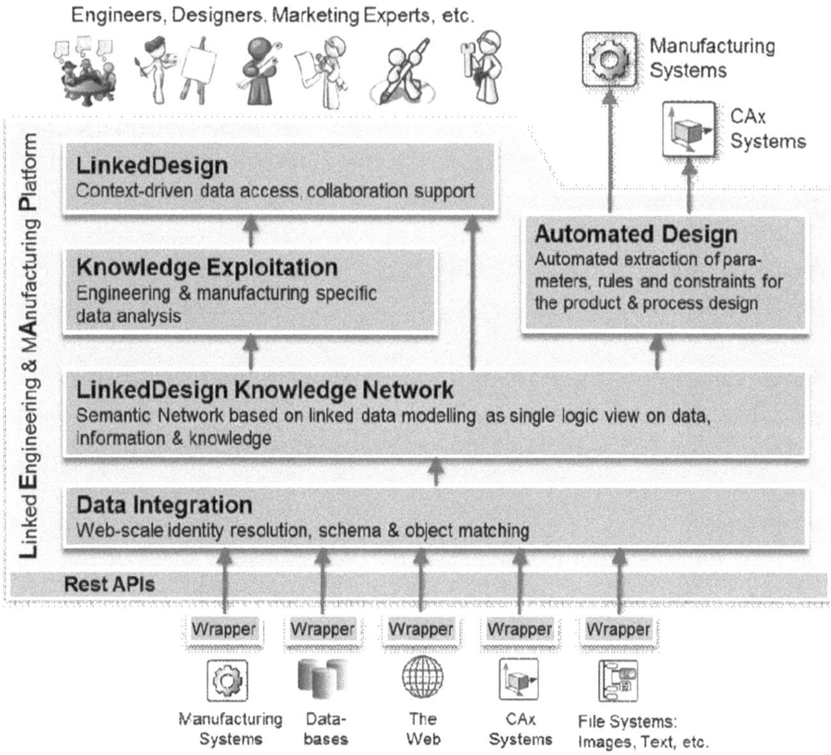

Figure 1.1 ***LEAP architecture.***

- The architecture design of the product, its main characteristics, and systems engineering constraints and requirements;
- The detailed drawings and product components (CAD designer);
- The design of manufacturing systems; and
- The evaluation and simulation of the performances of the product and the production process (analyst).

Second-tier users: These are users who will benefit from the LinkedDesign engineering platform using a more direct and immediate link to product and process design documentation. Details on the product and process design are directly exploitable during the execution of manufacturing processes that are controlled by Manufacturing Execution Systems (MES). These users are mainly production engineers, who will not only benefit from the closer link to the LinkedDesign embedded design knowledge but also contribute to it with their feedback and hints, thus closing the information loop between the design and production stages. This loop will also be enriched with production data from the MES

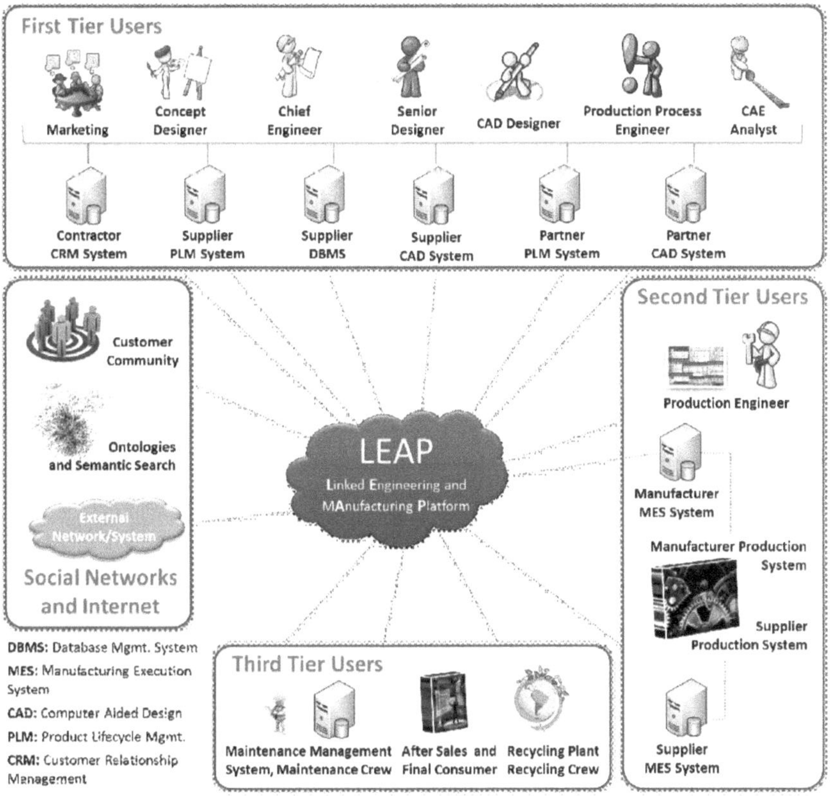

Figure 1.2 ***Users of LEAP.***

system, which will provide a better basis for product or process design improvements.

Third tier users: These users, even if traditionally not considered by engineering platforms, have been taken into consideration by the LinkedDesign system since they can benefit from the engineering information contained in the system to improve maintenance and dismantling (due to their direct link to design knowledge), as well as making a valuable contribution through their feedback and field information.

Finally, within the LinkedDesign application scenarios we focus on not only on crossdomain but also intraorganizational information sharing (first and second tier users), while the integration of third tier users requires interorganizational data security and privacy agreements. For LinkedDesign developments, we assume that all data security and data access authorization aspects are predefined and can be modeled within user roles that control the context-driven data access.

REFERENCES

[1] http://ec.europa.eu/enterprise/policies/industrial-competitiveness/economic-crisis/files/eu_manufacturing_challenges_and_opportunities_en.pdf.
[2] OECD. Policy responses to the economic crisis: investing in innovation for long-term growth. OECD Report; June 2009. Available from: http://www.oecd.org/dataoecd/59/45/42983414.pdf.
[3] Digital Agenda; 2010. Available from: http://ec.europa.eu/information_social/digital-agenda/index_en.htm.
[4] European Union; 2020. Available from: http://ec.europa.eu/eu2020/index_en.htm.
[5] Watson Healy L. Information industry outlook 2008. Report from Outsell; 2008.
[6] Feldman S, Sherman C. The high cost of not finding information. An IDC White Paper; 2001.
[7] Teevan J, Alvarado C, Ackerman MS, Karger DR. The perfect search engine is not enough: a study of orienteering behavior in directed search. In: Proceedings of the SIGCHI Conference on human factors in computing systems; 2004.
[8] Open Data Protocol, OData, http://www.odata.org/.

CHAPTER 2

LEAP Data and Knowledge Integration Infrastructure

E. Peukert and C. Wartner
University of Leipzig, Leipzig, Germany

Contents

2.1 OVERVIEW

Design, engineering, and manufacturing data is often distributed across a variety of heterogeneous systems. Relevant information about a product is scattered in different formats and locations, and it is tedious and error prone to search for information across different tools and data formats. Finally, information about a product is increasingly not being stored in one of the specific structured document formats (such as relational data, XML data, Excel files, CAx files, etc.) of specific tools, but is hidden in

Taking the LEAP
http://dx.doi.org/10.1016/B978-0-12-805263-1.00002-5

unstructured formats such as (the text of) office documents, emails, or enterprise-internal forums and blogs. In fact, the ratio of unstructured data amongst all data is estimated to be 80–85%, leaving structured data far behind in second place. In summary, information resides in a huge number of disconnected information silos, and the user might not have genuine access or—even worse—might not be aware of these silos. Thus, a single point of entry for all relevant product information—independent of its format, location or originator and the phase of the product lifecycle in which the information is created—is of paramount importance. Moreover, implicit (new) information which could be derived from existing information by embracing its mutual relationships and interdependencies remains invisible.

Data federation is one of the main goals of LEAP, it is also a prerequisite for all of the other goals. LEAP aims to integrate all of the relevant information across trusted sources in the product lifecycle, independent of its format, location, and origination time.

The data integration layer of LEAP consists of:

- The Smart Link Storage as a bottom layer persistency service for all LEAP components. The Smart Link Store is a graph store where objects imported from various data sources (or reference to them) and the relations between them are stored to enable an integrated view of all of the relevant data sources in the product lifecycle for analysis, search, and visualization purposes.
- The semantic mediator and data source wrapper components that allow us to access all kinds of underlying, application specific data sources in a uniform way.
- The matching framework providing schema and object matching approaches. The aim of the framework is to ensure the alignment of data sources, both on the meta and the instance level, and identify new links between information objects.
- Tools that support the analysis of the integrated and connected graph data.

These service-based tools can be used individually by any component of LEAP or they may be used external to LEAP, and graphical user interfaces exist for them. However, the services are designed to be integrated and form a dedicated infrastructure for data integration and analysis: The LEAP Smart Link Infrastructure shown in Fig. 2.1.

The presented tools will be described in more detail in the following sections.

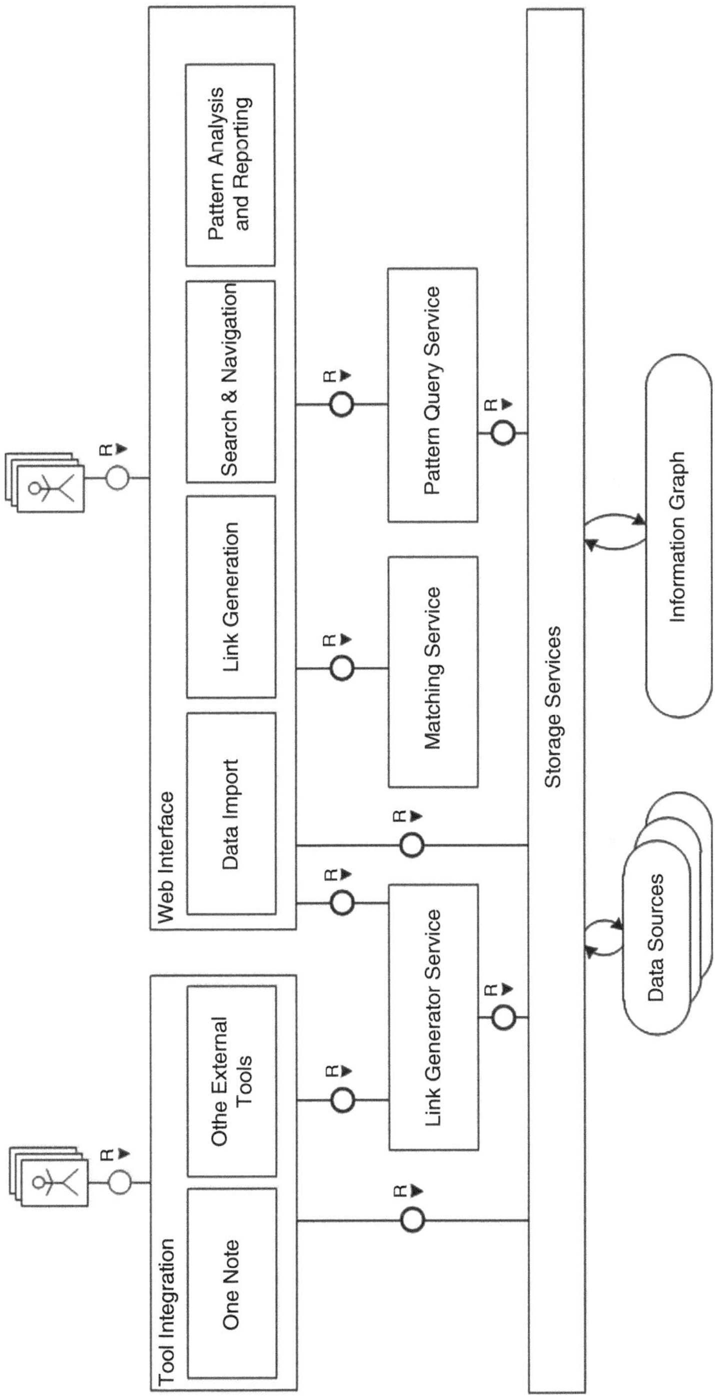

Figure 2.1 ***Smart link infrastructure.***

2.2 SMART LINK GRAPH

2.2.1 Motivation

One of the goals of LinkedDesign is to ease information federation, which is relevant to the full product lifecycle and thus to the support for the process of finding relevant information. In this context we propose the idea of a smart link storage—a collection of links (relations) between different kinds of entities. In LinkedDesign we concentrate on the diverse engineering and manufacturing processes of the product lifecycle, which consists of different entities. Examples of such entities are data elements, such as design documents, text fragments, process steps, tagged instances, concepts of ontologies, and so on. Also not to be neglected as parts of the overall processes are the "human" entities, such as experts in the field who have specific knowledge concerning the data elements. Links in this sense describe the relationship between entities. The links add knowledge to the platform and allow different deductions to be made, as well as reasoning and knowledge exploitation.

Saving and exploring links between connected entities can be used in many ways. We have already mentioned their advantage for knowledge exploitation, but further topics such as connection graphs can be exploited to help identify different patterns of relationships between entities. The net of links can also be used for tracing. Based on the graph (if entities are vertices and links are connections, the net of links can be seen as a graph structure), different analytics can be applied over the connected entities. In summary, having links between different entities improves the data federation and facilitates other "smart" services to extract knowledge and provide relevant information.

2.2.2 Concepts

First of all, we have to ask how to store links efficiently and at the same time remain generic. In the context of LinkedDesign, where the major topic is the product lifecycle, it is important to have a certain scope for the data that will give the answer to a user question. Thus, only a certain scope of data should be maintained, for example, data should be filtered depending on the context. For this reason, we propose to store relevant links using a database system which will maintain only the relevant data and will not try to link all of the available data in the world (which in this approach will not scale). Nevertheless, we expect a high load of data and queries over stored links. For this purpose we plan to use an in-memory database, which can handle such load and at the same time has very low response times. Additionally, to adhere to the overall LEAP architecture and to provide all the advantages that a service architecture offers, we will expose this functionality (store and load) as REST services.

A second topic that should be addressed is the ability to store probability values for the links, which describe how trustworthy the link is. The ideas of linked data assume that new data will be manually linked to other existing data. However, this "linkage" is a tedious process, which consumes a lot of time if performed manually. Consequently, the idea of creating the links automatically was developed [1,2]. Nevertheless, automatic approaches do not provide a hundred percent certainty that a created link is correct. This leads to the core idea of the smart link storage—to save the certainty value of the link together with the link itself.

We propose to store links using the following "quadruple" format:

<source, target, confidence value, link type>

This format has the following elements:

- *Source*—an URI identifier of the source entity.
- *Target*—an URI identifier of the target entity.
- *Confidence value*—a float value representing the confidence of the created link. It can be either computed, if generated automatically, or it can be a specified by the user. The value should be in the interval [0.00, 1.00]. The lower value represents a lower confidence and a higher value represents a higher confidence that the link is correct. The value 1 represents the full certainty that a link is correct.
- *Link type*—a description of the stored link.

The proposed quadruple format can be used to store links that were automatically created or user-generated together with their confidence values. This option directly relates to the schema and object matching services that automatically find similarities between certain types of entities. Thus, the smart link storage can also be used as storage for the results of the schema and object matching services.

The quadruple concept gives different opportunities for applications to exploit the stored net of links. Applications can query and interpret the links to provide a more thorough knowledge exploitation of the available data. Furthermore, depending on the confidence value of the involved links, a certain confidence of the answer to user queries can be given. To give an example, let us assume that we have the following links:

- <document1, specification1, 1.00, is related to>
- <specification1, law1, 0.92, is related to>

The first link represents a link of type "is related to" from the source (a document in this case) identified by the URI "document1" to the target identified by the URI "specification2" (a specification document). The second link represents a link of the same type "is related to" from the source entity identified by the URI "specification1" (thus, it is the same specification document from the other link) to the target entity identified by the URI "law1" (eg, a link to a public web site where the law has been published). The smart link storage will offer both store and load services for such links. More detail about the infrastructure and the offered services is given in Section 2.3.

From the stored links, an application can not only load the saved links and present them as data to the user, but it can also derive the knowledge that document1 is related not only to specification1 but also to law1 by following the graph of the links (document1 relates to specification1, which on its side relates to law1). Thus, the "smart" application, built upon the smart link storage can present the user with the "explored" knowledge that document1 relates to law1. The user can also see with what confidence this knowledge is considered to be trustworthy. Different calculation scenarios for the accumulated confidence value can be applied. For example, some of the options include:

- Multiplication of confidence values over the graph path of the explored links—$1.0 \times 0.92 = 0.92$ confidence value of the derived link.
- Mean value—$(1.0 + 0.92)/2 = 0.96$ confidence value of the derived link.
- Minimum (0.92) or maximum (1.0) confidence values on the path.

2.2.3 Smart Link Service

The smart link storage is exposed by offering several REST services to store and query data, as illustrated in Fig. 2.2.

The Smart Link Service is the core component of the Smart Link Infrastructure and it allows us to retrieve specific sets of nodes from the graph based on a pattern specification.

The Smart Link Service relies on a property graph model similar to that proposed in Ref. [3]. A node has a set of properties. Properties can be associations to other nodes or scalar values of simple types.

Each node has a type which can be seen in analogy to a table in a database schema. However, due to the flexibility of the model, the type structure (the collection of possible properties) is not fixed but can be derived from the actual instances of a type. For example, it could happen that a user

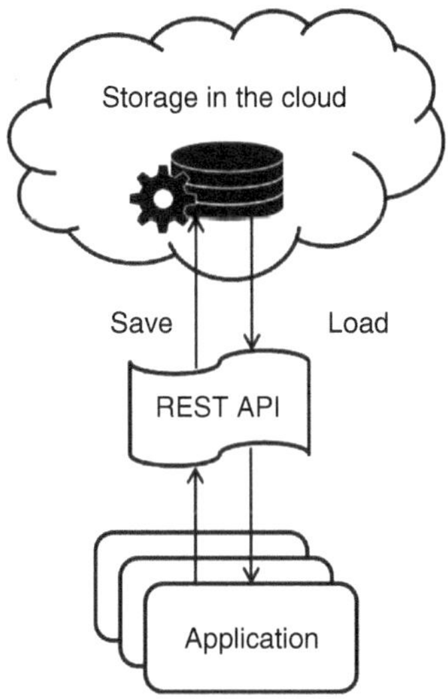

Figure 2.2 ***Smart link storage.***

imports a list of employees and stores their *age* and *name*. The imported employees are typed as *person*. The type *person* does not need to exist and will be created on request. If another user also imports employees, he or she can add an additional property *income* without even being forced to change the existing schema/type person. The graph store now allows us to dynamically compute the schema which specifies that the type *person* has three properties. It is also possible to measure how many instances actually carry such a property.

In addition to nodes, we can import links. Links are also nodes, but with a predefined set of properties, that is, a quadruple that stores the source, target, link type, and confidence of a link.

2.3 SOURCE WRAPPERS/MEDIATOR

2.3.1 Motivation

Current business IT is mainly characterized by a heterogeneous IT infrastructure. In other words, a broad variety of different systems and applications are available to achieve an adequate management of data, as well as provide business related information to the users. Examples include CAx tools, simulation or office-applications, other systems like wikis, search engines, content management systems, or product data management systems. These systems and applications are typically used because they allow quick access to documents, CAD models (or drawings), production plans, reports, guidelines, and similar documents.

In this context, the concept of a database management system has been adapted to different needs and it appears in various forms in order to manage different types of information. For instance, product requirements and consumer needs are managed by requirements management software. CAD models and design variations are managed by product data management software (PDM). Documents, rules and guidelines are managed by document management systems (DMS). FEM-data (FEM—Finite Element Methods) and models are managed by simulation management software and unstructured information snippets, for example, a collection of best practices, are stored and managed by a wiki-like software or an intranet portal.

All of these information management systems have special features, such as to manage approval processes or ensure conformance to predefined workflows. All of these systems are suited for special production purposes and, therefore, their coexistence is useful and intended. However, the consequences of this trend are that the employees are required to use several information management systems in parallel. To get fast access to the necessary company specific information, a user has to use an increasing amount of implicit knowledge about the company's specific possibilities on information retrieval, for example:

- What kind of company specific information systems and repositories exist?
- What kind of information is stored in the respective repository?
- What kind of document metainformation is explicitly stored and, thus, applicable for an intelligent search?
- What language has to be used in the specific repository or by its search engine?
- What numbering systems, taxonomies, abbreviations, or acronyms are used?

Despite the user's skills, which are increasingly required, data and information exchange between systems is still a difficult challenge. Although nearly all systems provide functions to import and export data, integration is difficult to be reached because most of the systems are based on application-specific standards and interfaces, making application-independent queries or data-filtering all but unfeasible.

In theory, standards like Extensible Markup Language (XML) or Web-Services support a system-independent storage and exchange of data, as well as a remote use of application-specific features. But a "real" data-integration is not guaranteed this way because the effort is simply outsourced to a transformation script. Because of missing coherent semantics, a consistent and logical exchange, even between only two different systems (eg, an ERP and a PLM system), still implies an extensive mapping effort. For example, someone has to know that a "time stamp" in system A is identical to "status time" in system B.

In order to enable an integration of heterogeneous data-sources, a common exchange-format or a common level of abstraction is required, which provides coherence even on a semantic level. A context-related use of data from different domains will become feasible only by enhancing data with semantic information (ie, metainformation).

2.3.2 Concept

The Semantic Mediator is a LEAP module, which enables read and write access to arbitrary data sources via a bunch of file-related (eg, *.CSV) or database-related (eg, SQL) system-interfaces. Queries processed through the Semantic Mediator are transformed into subqueries, each in accordance to the constraints given by the specific repositories and adapted to each repository. This allows a sent request to be processed by all linked sources in one single step. The delivered query result contains no duplicates or inconsistencies.

The core idea of the Semantic Mediator implies a semantic description of the linked data sources. This description is realized by ontologies, which provide the specification of the accessible entities as well as their semantic relations. This enables the solution to solve heterogeneity conflicts. Ontologies are able to build a common semantic basis, which enables a coherent representation of data and related data sources. According to Ref. [4], a semantic model is characterized by the following three properties:

- common concepts;
- relations between objects;
- description logic.

In principle an ontology can consist of one big global or several small subontologies, which can be merged at runtime [5]. This approach includes the idea of having several distributed ontologies, each representing a data source. The ontologies are federated through an "Inter-Ontology-Mapping" that is based on a specific query. This means, that the mediator generates a common ontology consisting of X subontologies for X data sources which are affected by the query. This approach leads to better scalability as well as flexibility because every information system has its own ontology and new sources can be linked by adding new ontologies, even if the semantic mediator is already in use.

Accordingly, the system architecture is designed in a modular way. Incorporating a particular data source via the Semantic Mediator is implemented via so-called wrapper components, which are linked to the core Semantic Mediator module over a standardized interface. The wrapper components encapsulate all of the elements, which are necessary for the integration of new data sources. In particular, they contain the semantic descriptions of the repositories, as well as the corresponding transformation mechanisms.

2.3.3 Source Wrappers

As outlined previously, different kinds of data-sources can be linked to the Semantic Mediator simply by implementing a data wrapper. In the following, the principle structure as well as the implemented/specified wrappers are outlined.

The complete information, which is required for the communication with a specific data source is encapsulated in wrappers. They contain a semantic description of the respective data source in terms of an ontology as well as a transformation mechanism between the global semantic description (mediator ontology) and the internal semantic description of the data source (eg, ER-schema of a database). In most cases, these transformation mechanisms are implemented as simple mapping files (XML based). Such files describe the mapping between the entities of the ontology and the structural elements of the linked data source. For example, the column-name of a database table corresponds to an entity of the ontology in a SQL wrapper.

It can be deduced from this approach that each data source needs its own wrapper, or more concretely: two SQL databases need two SQL wrapper instances. This is necessary because each of the databases contains its own semantics. Even though both mappings can rely on an SQL wrapper, two instances of this wrapper have to be deployed. Each one contains a subontology, which describes the semantics of respective data source, and a mapping file. Hence, the process of linking a data source can be split into two kinds of use cases: linking a data source, which corresponds to a known type, and linking a data source, which corresponds to a "new" type.

For the linkage of known data source types (eg, SQL) new instances of wrappers are deployed by copying a wrapper template into the used wrappers section and configuring the files according to the given semantics. For new types (eg, if you want to add a non-SQL database) a new wrapper type has to be developed. Thus, wrappers can be fundamentally different in their design for linking the data into the Semantic Mediator (depending on the heterogeneity of the data sources).

Significant effort is required to implement several wrappers just to be able to link several files to the Semantic Mediator. Thus, wrappers aiming to incorporate files (eg, csv or xls) are designed to provide access to several files at once by incorporating the precondition that all files are common in their structure. In other words, if the files are semantically coherent, then there is no need to have several instances of such a wrapper. If not, then there is currently no workaround. Currently, there is no established alternative on the market for reliable externalization of the domain specific semantics aside from manual work. However, the schema matching services may be capable of providing a solution for an automated mapping in this context.

As described previously, different wrapper types are necessary to link different types of data sources. Each reflects the characteristics of a specific data source. With respect to the LinkedDesign challenges, currently four different kinds of wrappers are in focus:

- SQL-Wrapper
- CSV-Wrapper
- XLS-Wrapper
- XML-Wrapper

Even though not all wrappers are in a ready to use mode, sample implementations can validate the concept of linking heterogeneous data sources into one single interface.

2.4 DATA MATCHING AND LINKING

2.4.1 Motivation

To successfully reach LEAPs goal of federated data, which is to be able to holistically analyze and search within data that is distributed across multiple heterogeneous data sources, various data integration approaches are used. There are some similarities to ETL (Extract Transform Load in Data Warehousing) processes that are used when building data warehouses to allow efficient analytic processing, such as when data from independently created databases is to be combined to create a unified view that gives new insights for analysts.

The main tools that are used to support the data integration task are schema alignment and object matching. In LEAP both of these methods are provided through services, specifically the schema matching service and the link generator service. These web services are integral parts of the data integration infrastructure but can also be accessed by any client due to their standardized REST interfaces.

The general workflow of integrating data in smart link storage starts with the data import that is supported schema matching methods. In this step, the relevant data sources can be selected (eg, files, databases, etc.) and a mapping of their schema with the schema of the Smart Link Store is created. Based on the mapping, new entities are added to the information graph. The use of the schema matching service is optional and supports the administrators during the definition of the mapping. However, it can simplify subsequent steps if a sound input schema is available. In the linking step, object matching approaches are applied to identify the relationships between entity sets that originate from different sources that often contain no direct references to each other. In contrast to standard link approaches, we also support text mining techniques and new mapping types. Text mining helps to deal with unstructured data, such as text documents, whereas new mapping types are needed to determine the type of semantic relations between entities.

A number of approaches can be found in literature that simplify the creation of federated data sources [SL90, BLN86], that use schema matching Ref. [6,7], or that apply object matching to identify links between entities [8,9]. However, most of the existing work only focuses on individual aspects of matching and often come with the restriction of fixed schemas, whereas the LEAP data integration infrastructure aims to provide a whole set of tools that support every step in the creation of an information graph, connecting vastly heterogeneous data sources in design, engineering, and manufacturing.

2.4.2 Schema Matching Service

Schema Matching aims to create mappings between data base schemas or other metadata descriptions during the import of new sources to ensure a consistent global schema. Fig. 2.3 shows a mapping between two different purchase order schemas. This is crucial for later operations because they then can assume that data objects of the same type have the same metadata.

This is highly important when new information is imported to the smart link storage. The store is schema less and allows objects of any type with arbitrary attributes, even for objects of the same type. Even though no common schema is enforced by the smart link, the storage analysis operation needs aligned object metadata to work effectively. For example, the

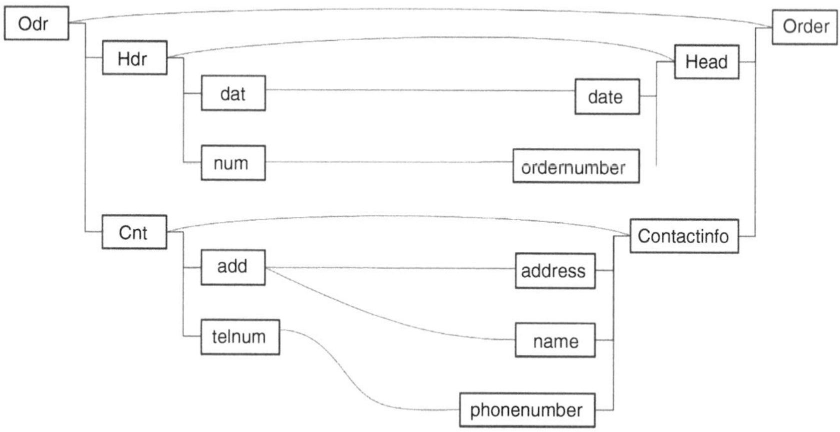

Figure 2.3 ***Mapping between object metadata.***

task of finding information about persons of a certain age is much easier and efficient by querying all objects with the type Person and an 'age > 25' instead of multiple queries for persons with a certain 'date of birth', 'age' or 'birthday', and so on.

The Schema Matching Service semiautomatically computes a mapping suggestion for the user. The matching service contains a number of matching algorithms and a library of schema importers for different schema types [10,11]. It takes two schemas as input and computes a mapping suggestion between them. Similarities between source and target elements are computed not only on the metadata level but also on the instance level. Since current matching systems are often not robust enough to be able to cope with very heterogeneous source schemas, we have developed an adaptive matching approach [11]. This approach automatically configures a schema matching system process that consists of a set of operators for matching and filtering. The so-called matching rules can be defined based on the measured features of the input schemas and the intermediate results. These rewrite rules rely on analyzing the input schemas and intermediate results while executing a process and rewrite the process to better fit to the problem at hand.

Since the handling of the two services and the HTML pages is quite cumbersome for testing, a dedicated schema and mapping visualization was developed that can be used to perform automatic matching. A screenshot is shown in Fig. 2.4.

This tool allows a user to import a source and a target XSD schema, and is able to suggest a mapping. SVG and HTML5 technologies are used

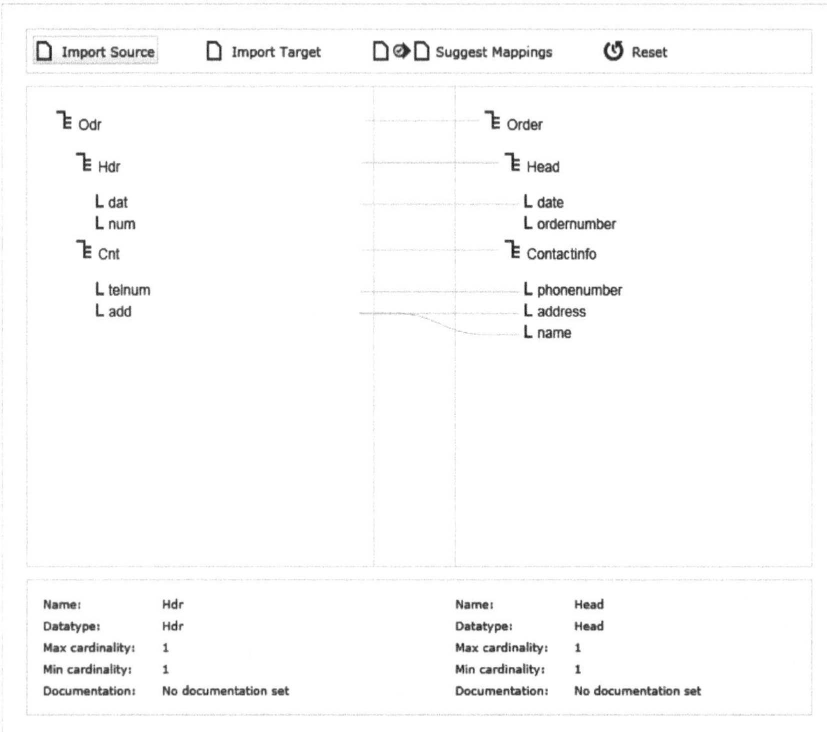

Figure 2.4 ***Visual Mapping UI.***

to draw the lines. This tool can be integrated in the LinkedDesign Virtual Obeya and is used in the data import tool.

2.4.3 Link Generation

In contrast to mappings on the metadata level, Object Matching approaches work on instance data. They take a set or multiple sets of objects as input and calculate the confidence of a semantic relationship (eg, similarity in the common use case of duplicate detection [12]) for each object pair. It then returns a mapping that consists of a set of correspondences. Each correspondence contains an object pair, their semantic relation, and a numeric confidence value between 0 (uncertain) and 1 (certain) for this relation. Thus, the correspondences are in fact links between objects. Fig. 2.5 shows a list of records of people with duplicates. However, the deduplication process can be cumbersome due to typos, abbreviations, and different representations.

Creating graphs of linked objects from a structured data source such as a database with well-modeled metadata can be relatively easy. In contrast,

LastName	FirstName	Street	City	Gender
Smith	Kristen	2 Hurley Pl	South Fork, MN	female
Smith	Christian	Hurley St 2	S Fork MN	male
Smith	Christoph	23 Harley St	Chicago, IL	male
Smith	Kris L.	Hurley Place 2	South Fork, MN	female

Figure 2.5 ***Mapping between two instances referencing the same person.***

creating a graph from unstructured sources like document collections or independently created databases requires a component to find links between entities. The Link Generator Service allows the creation, management, and execution of workflows to determine relations between entities, the relation type and the confidence of these relations. These workflows are related to the object matching workflows that are used in various data integration and data quality related scenarios. Fig. 2.6 shows how matching workflows look at an abstract level. Normally, they are used to detect data objects that are equivalent in the real world but have different representations in multiple data sources.

The inputs of an entity resolution workflow are usually sets of data records from one or multiple databases. The operations in the preprocessing step include data transformation steps (eg, to convert data types or remove special characters) and filter operations or blocking steps to reduce the search space for finding matching objects. Match operations are then applied to determine the pairwise similarity of the candidates. Depending on the domain, different distance metrics on one attribute or a combination of attributes can be used to determine the likelihood of two entities being equal. The match result usually is an instance-level mapping: a set of correspondences of the type (*entity1, entity2, sim*) where *sim* is the confidence of two entities being equal.

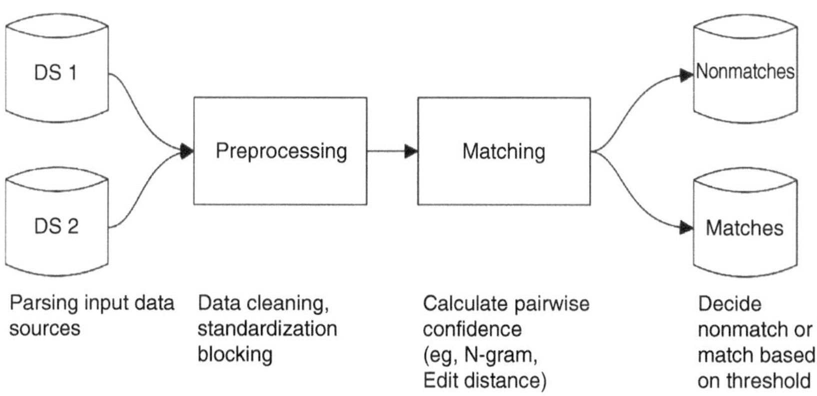

Figure 2.6 ***Classic matching workflow.***

Standard entity resolution workflows can be used to find objects in the graph store that are very similar and then merge them under the assumption that these are duplicate nodes or slightly different versions of the same note. However, these data quality related workflows are not the main focus of the Link Generator service in LEAP.

Instead of searching for pairs with a high confidence of a *'sameAs'* relation, linking workflows aim to search for links between arbitrary objects and their type. In the easiest case, a linking workflow is very similar to a simple matching workflow and just determines a link between two objects by calculating the similarity between certain fields in two objects. The following example is a workflow that tries to find documents written by the same author:

1. Parse the Input (eg, a document with a metadata field about the author).
2. *Preprocessing*—Split the author field into individual names and sort them alphabetically.
3. *Matching*—Calculate the trigram similarity of the sorted author list.
4. The result is a set of correspondences between documents that in effect have the type *'sameAuthor'*.

Linking workflows like this are useful to generate certain types of edges for the property graph that connect objects of the same type. However, to enable the functionality needed for the Smart Link Infrastructure, that is, the calculation of links between arbitrary data objects representing entities like persons, documents, etc., most of the steps depicted in Fig. 2.7 are extended and an additional, optional, rule application step is added to the process. Fig. 2.7 shows a general linking workflow.

Especially when trying to support search and information discovery in knowledge management processes, the data often consists of unstructured text from documents with possibly incomplete metadata.

The string similarity of the content of two text documents, for example, does not provide enough information about the relationship between them.

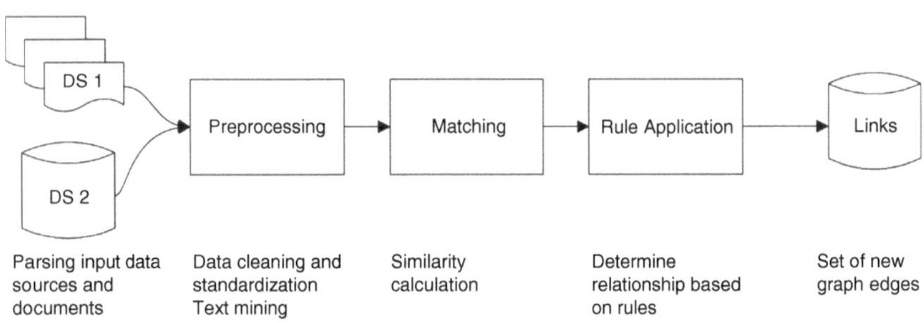

Figure 2.7 ***General linking workflow.***

The only relationship that could be derived from a low lexicographic distance is that the documents have duplicate text or large parts of overlapping text. Named entity recognition, keyword detection, the detection of hyperlinks and mail addresses as well as other text mining approaches become important to link documents with semantic relations that go beyond *'sameAs'* relations.

A simple linking workflow that relies on text mining operations and which uses a simple rule to decide the relation type could look like this:

1. Parsing of Input data (eg, the text of a K-Brief).
2. *Preprocessing*—Named entity recognition, Keyword detection.
3. Calculation of the similarity (eg, amount of overlap) of the determined list of keywords, engineering concepts.
4. *Application of rules*—For example, a high overlap of keywords leads to the relation *'similarContent'* and a high overlap of engineering concepts extracted from the two documents lead to the relation *'sameTopic'*.
5. The result is a mapping containing correspondences of the type *'similarContent'* and *'sameTopic'*, that is, links that are inserted into the graph.

In addition to the use of text mining approaches, linking workflows can also utilize existing edges in the information graph. An example would be a relation such as *'sameTopic'*, which is transitive to some degree. Once a linker determines that a new and an already integrated document have a high confidence of such a relation, we can infer that the given relations exist between the new document and the already integrated documents neighborhood. This is similar to data matching approaches that use already existing mappings [13] and can be useful to discover and verify certain types of relations, as well as to increase the efficiency of a linking workflow.

Generally, the workflows of the Link Generator service first produce mappings with instance-level correspondences of the form *(entity1, entity2, linking method, confidence)* representing the used linking method and its confidence [eg, *(document1, document2, trigram-author, 0.8)*, *(document1, document2, keywordoverlap-text, 0.2)*]. A final rule application step decides what edges are added to the graph based on the applied methods and their confidence value.

The Link Generator service offers a variety of preprocessing, text mining, data matching approaches and filtering operations that can be freely combined. These include:

- preprocessing operations, filter operations to exclude certain entities from subsequent operations;
- keyword, link, address and named entity detection operations that enrich objects with attributes containing lists of entities and other information;

- dictionary based unification of terms;
- match approaches like N-gram distance, TF-IDF distance, Edit distance to calculate the confidence of attribute similarity (an overview over these common metrics can be found in [13]);
- match approaches that determine a confidence value based on the overlap or element similarity of lists;
- match approaches based on already known relations, that is, the neighborhood of entities [5];
- combination operations for aggregating confidence values determined by different approaches;
- relationship detection rules based on individual or combined confidence values.

Even though a number of linking workflows were created, a fixed set of available workflows would limit the usefulness of the Linking Service. Despite the existence of some operations that are useful in any situation, the ability to extend the service with new linking workflows is important to achieve the best results depending on data sources, application domain, amount of data, and the planned analysis operations.

The operation library of the Link Generator service offers a set of operations to be used. Each operation has a set of input parameters and can have three different types of results. The input parameters can be descriptions of data sources, thresholds for linking approaches, attributes to be processed, and so on. The result of an operation can be a set of instances (eg, preprocessing operations that annotate instances with new attributes like language detection operations), a mapping (eg, operations that detect links between instances), or a message (eg, operations that just stores the end result to a database at the end of a workflow and only confirm their success).

These operations can be differently parameterized and freely combined, which is limited only by the input and output type (eg, a mapping producing operation cannot be followed by an operation that tries to store a set of instances in a database, an operation that filters out links based on their confidence value must be preceded by a mapping producing operation, etc.).

Typically, a workflow for creating a set of links requires at least a step which defines the input, one operation, and finally a step for handling of the result.

Fig. 2.8 shows a series of workflow steps defined by a user for querying the Smart Link Store, applying a similarity metric to the query results, applying a filter and writing the resulting links back to the Smart Link Store.

After the creation, such workflows can be stored and executed periodically or triggered by calling a function of the Linking Service API.

Figure 2.8 *User defined linking workflow.*

2.5 SMART LINK TOOLS AND APPLICATIONS

2.5.1 Motivation

The goal of the Smart Link Infrastructure is not only providing individual components for data integration tasks but also to support a user during the whole process and to provide data analysis tools. Therefore, web based user interfaces were developed that improve usability and accessibility, especially for nontechnical users. The User Interface of the tools are web based. This allows the easiest integration with other tools either in LEAP (as a widget in the Virtual Obeya) or outside of LEAP.

2.5.2 Data Import Tool

The data import tool usually is the first step in the data integration and analysis workflow of the Smart Link Infrastructure with the exception of use cases where data is pushed directly in the Smart Link Store by external tools. The final version offers import possibilities for more data sources.

To prefill the Smart-Link Store with entities, an administrative user can use the Data Import Tool. Such prefilling allows a user to connect to existing data sources such as relational databases, Excel files, or data sources that are exposed through a SPARQL end-point. When integrating a data source into the information graph, the schema of the data source needs to be mapped to the existing flexible schema of the information graph. Such a mapping describes how elements of the source schema correspond to elements of the information graph schema. Although extending the graph schema is not problematic, existing types and properties should be reused. Adding properties to existing types or adding types is implicitly triggered when new entities are uploaded. Defining the mappings can be complex and time consuming and it is often done manually, with the help of point and click interfaces. The Smart Link Infrastructure data import tool offers a point and click interface that is similar to existing mapping solutions (Fig. 2.9).

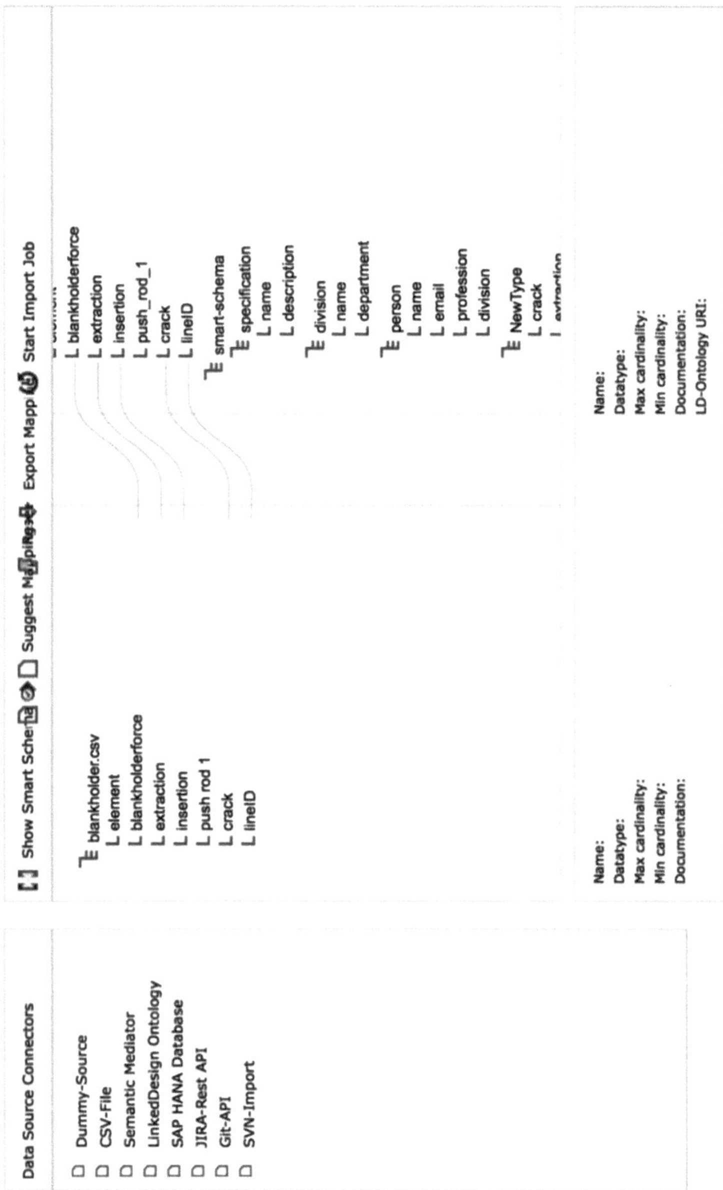

Figure 2.9 *Point and click data import tool.*

Moreover, a Schema Matching Service is integrated and enables us to compute automatic suggestions for mappings.

2.5.3 Linking Tool

Users can create, manage and test workflows via the GUI of Linking Tool. The workflow creation supports the user in creating workflows as a chain of steps that each require the selection of an operation and definition of parameters. Fig. 2.10 shows such a parameter selection. The results of individual operations or partial workflows can be previewed during the definition (Fig. 2.11). Stored workflow can then managed under the 'Manage' page that allows the workflows to be deleted, tested, and edited (Fig. 2.12).

2.5.4 Querying Graph Data

Pattern queries are a crucial part of analyzing process data holistically. Pattern queries are much more intuitive for a user than navigational queries since the user does not need to know the exact topology of the graph which is required for navigational queries. Navigational query access refers to the ability to traverse associations in a graph within the query language. Many graph stores only implement navigational queries such as GREMLIN in Neo4J

Figure 2.10 ***Workflow step definition.***

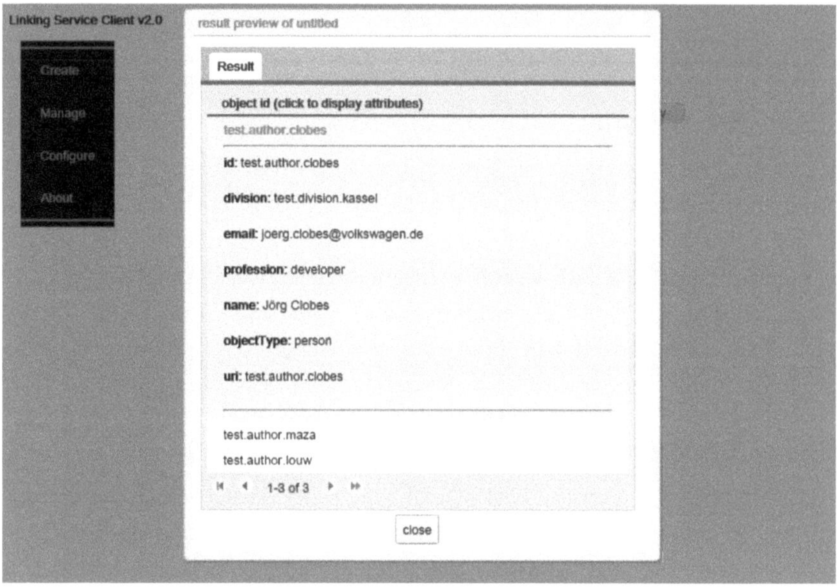

Figure 2.11 ***Preview of an object set.***

Figure 2.12 ***Management view of all created workflows.***

or WIPE within SAP HANA [3]. Existing pattern query languages include SPARQL for triple stores and Cypher for Neo4J property graphs. Due to the nature of triple stores, SPARQL queries often become rather complex. The Cypher query language for property graphs of Neo4J is much easier to use. Unfortunately, it is bound to Neo4J. Typically, the results to graph pattern queries are computed by applying maximum common subgraph algorithms, which are expensive to execute. Since we would like to rely on an existing internal property graph store that only provides navigational query access, we decided to implement a pattern query interface on top of the navigational query interface. A pattern query is taken as input from which a navigational query is generated. Fig. 2.13 illustrates the overall process of query generation.

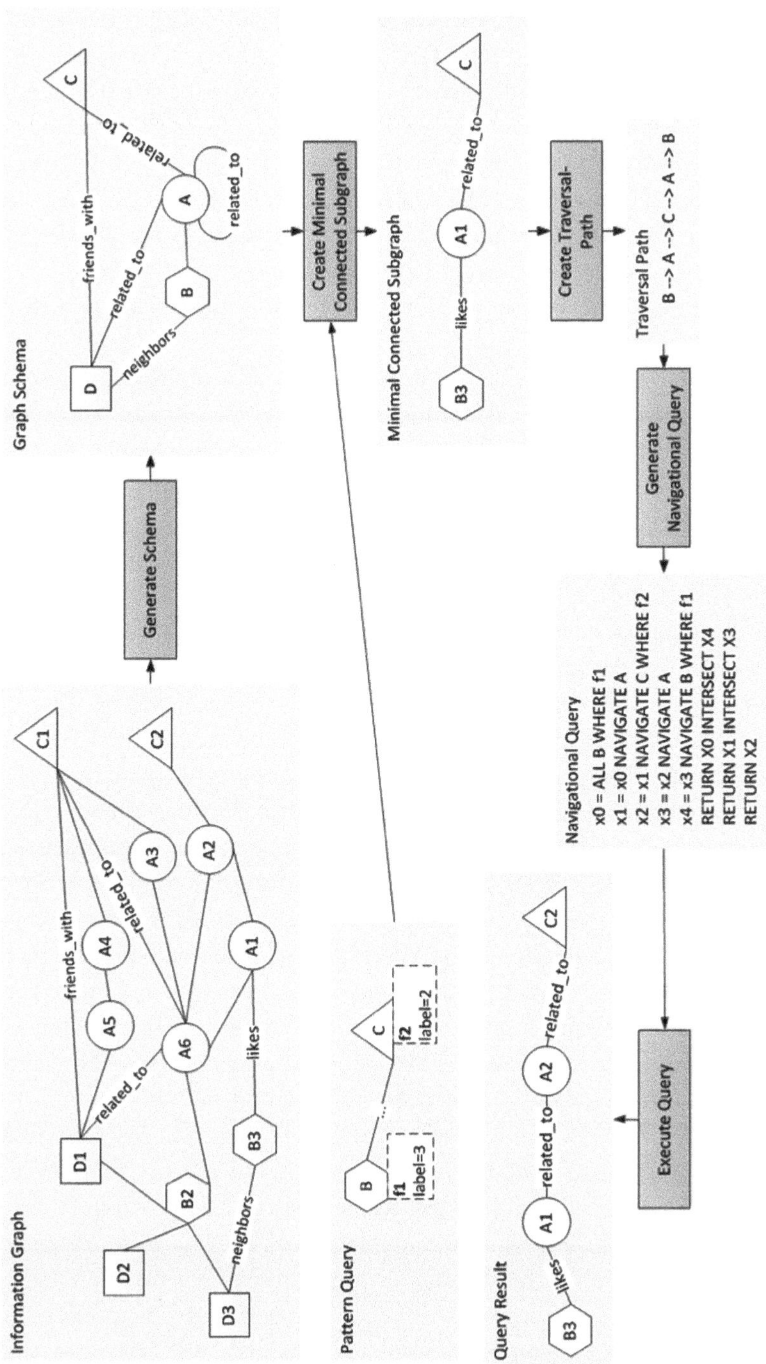

Figure 2.13 ***Pattern query generation process.***

As described earlier, the graph model that we use allows us to compute a schema dynamically from the instances that are currently in the graph. Before any query is issued to the graph store, the current schema is computed. This example assumes that there is a graph with four different kinds of entities, which are represented by different shapes and names of nodes. The generated schema only consists of four types of entities, including possible associations between instances of these types. A user could create a pattern query by stating that there shall be two nodes that are directly or indirectly connected and the nodes have to follow certain filter predicates. Such predicates define which nodes from the set of instances of a type should be in the result set.

From the schema graph that was initially created, a minimal connected subgraph for the given types is computed. If the types are not connected on the schema level, then the query will return an empty answer. In the example, the intermediate type A was added to get a minimal connected subgraph. To be able to generate a navigational query on that subgraph, a traversal path is needed based on the type graph. In the example, the traversal starts at B and walks to C. Note that the walk needs to be done in a backwards direction since initial filtering on the B nodes did not involve indirect filtering through the subset of C nodes. Consider that there are B nodes that follow the predicate f1 that have no possible path to any C node that follows predicate f2. After the traversal path has been computed, we can now generate the navigational query using any query language that allows us to navigate associations in both directions. This means that if there is an association from A to B, then the classical navigation from A to B is possible. However, in our approach we need to be able to navigate backwards from B to A, returning all nodes A that have an association to B. The query language used in the example is a simplified example. At the end of the query, the results of the individual navigation steps are intersected to return the correct set of nodes for each node type. In the example, nodes are returned that were originally not requested in the query. This can be filtered out depending on the requirements of the respective application.

Additionally OLAP-like operations can be expressed. We can now group along an attribute or node in the pattern query. In the example given in Fig. 2.14, we group along all of the KBriefs of the results that are computed by the pattern. There can be multiple groupings. For each group, an aggregation definition can be specified, which consists of a function and a binding to some node or attribute of the result specified by the pattern. In the example, two aggregation definitions are contained. One aggregates the

```
{
    "name": "queryName",
    "types": [
        {
            "uri": "com.sap.columbo.ws.kbrief",
            "id": "1",
            "included": true
        },
        {
            "uri": "com.sap.columbo.ws.rating",
            "id": "4",
            "included": false,
            "propFilter": [
                {
                    "prop": "com.sap.columbo.ws.value",
                    "comparison": "expression",
                    "value": ">0"
                }
            ]
        }
    ],
    "edges": [
        {
            "buri": "com.sap.columbo.ws.kbrief",
            "uri": "com.sap.columbo.ws.kbrief",
            "auri": "com.sap.columbo.ws.rating"
        }
    ],
    "aggregation": [
        {
            "bindings": [
                {
                    "typeURI": "com.sap.columbo.ws.rating",
                    "propURI": "com.sap.columbo.ws.value",
                    "id": 0,
                    "alias": "$type1"
                }
            ],
            "function": "AVG($type1)",
            "functionExternal": "AVG(rating/value)"
        },
        {
            "bindings": [
                {
                    "typeURI": "com.sap.columbo.ws.rating",
                    "id": 0,
                    "alias": "$type2"
                }
            ],
            "function": "COUNT($type2)",
            "functionExternal": "COUNT(rating)"
        }
    ],
    "grouping": [
        {
            "typeURI": "com.sap.columbo.ws.kbrief"
        }
    ]
}
```

Figure 2.14 *Pattern query example.*

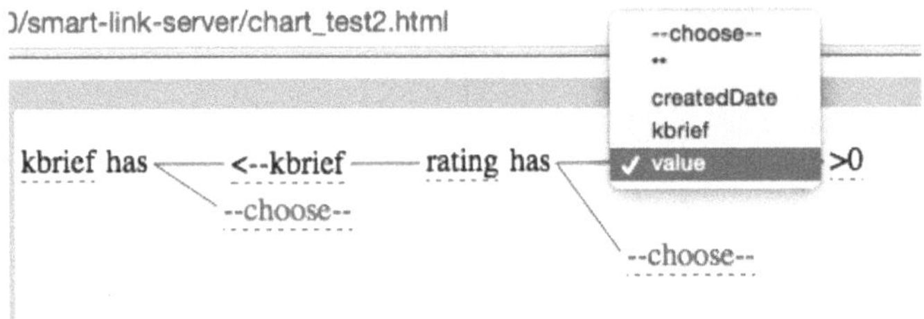

Figure 2.15 ***Pattern query UI.***

values of ratings by building the average whereas the other definition counts the number of ratings within each defined group.

This basic extension of the pattern query API allows us to express a number of analytical queries on the graph. The key point is that analytical queries are expressed on top a pattern query, which is different to typical OLAP. The pattern query serves as a projection and filtering step, whereas the aggregation and grouping define the actual analysis.

Obviously, the defined JSON-Schema is rather complex. Therefore, a graphical component that allows us to model such queries without knowing the actual query model was developed. Fig. 2.15 shows this UI. The UI is highly interactive in that users can click on any node and select available properties.

It serves as the metaphor to specify which data of a graph is of interest when building analytics.

2.5.5 Search and Analysis

The Smart Link Infrastructure offers tools for search and analysis within the integrated information graph. The information graph with its links can help to improve the ranking (see PageRank from Google) of a keyword based search result. Moreover, the graph that can be found in the vicinity of a node is displayed in addition to the search result and entities can be filtered by facets. A screenshot of the developed search interface can be found in Fig. 2.16.

The analysis of the information graph is driven by pattern queries and a simple drag and drop metaphor for creating charts. Fig. 2.17 shows the charting tool. It enables a user to graphically create pattern queries. A new report can be created by dragging and dropping dimensions and measures onto the axes of an empty chart. In addition, users can define functions over measures to compute complex measures on top of the pattern query result. Additionally the user can edit the aggregation functions that are used for

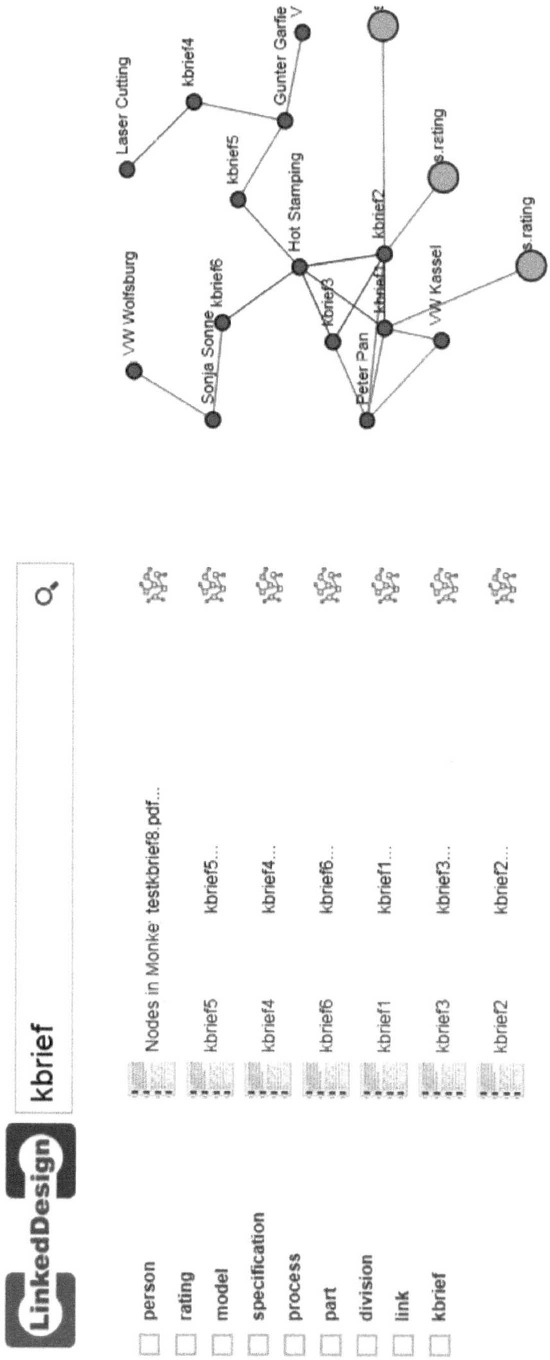

Figure 2.16 ***Search UI.***

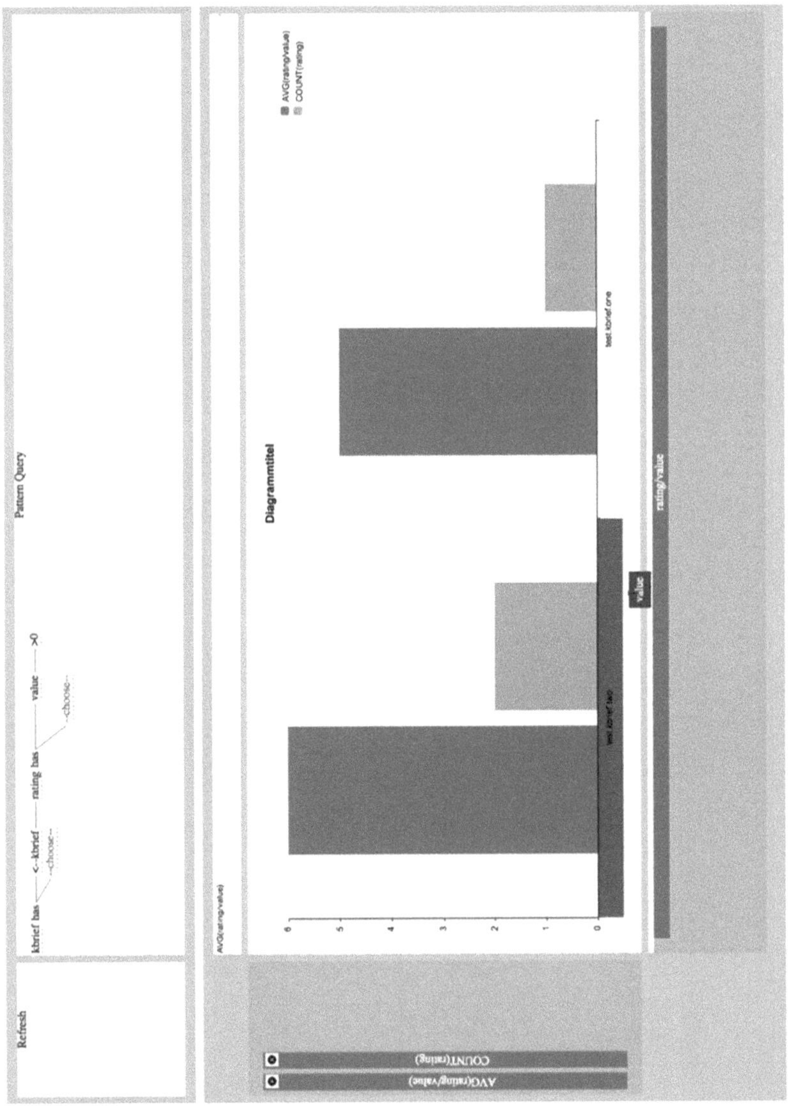

Figure 2.17 *Charting Tool—drag and drop to X- and Y-Axis.*

the Y-Axis. These functions can be complex since adding or multiplying the results of predefined aggregation functions is allowed.

2.6 CONCLUSIONS

This chapter has introduced the LinkedDesign vision and derived the requirements for a LinkedDesign conceptual architecture. The conceptual architecture of LEAP with its three-layer approach was introduced. We can conclude that the initial vision was very broad in that it tried to cover many very heterogeneous aspects together in one platform. The outcome is a conceptual architecture that can be instantiated be different use-cases. This was done in the LinkedDesign project for three very different manufacturing and design use-cases.

A basic part of the LEAP-Platform is that its data and knowledge integration infrastructure allows us to bring together heterogeneous data sources from a complete product lifecycle to be able to perform integrated analysis. This infrastructure is a key component of LEAP, which provides most opportunities for reuse within other knowledge integration projects of a similar scale.

REFERENCES

[1] Volz J, Bizer C, Gaedke M, Kobilarov G. Discovering and maintaining links on the Web of Data. In: Proceedings of the eighth International Semantic Web Conference ISWC09; 2009.
[2] Scharffe F, Euzenat J. Linked data meets ontology matching—enhancing data linking through ontology alignments. In: International conference on Knowledge Engineering and Ontology Development, KEOD; 2011.
[3] Rudolf M, Paradies M, Bornhövd C, Lehner W. The graph story of the SAP HANA Database. Magdeburg: BTW; 2013.
[4] Maedche A, Motik B. Repräsentations- und Anfragesprachen für Ontologien—Eine Übersicht. In: Datenbank-Spektrum (2003), Nr. 6, S. 43–53.
[5] Wache H. Semantische Mediation für heterogene Informationsquellen. Berlin, 2003.
[6] Bellahsene Z, Bonifati A, Rahm E. Schema matching and mapping. Heidelberg, Springer; 2011.
[7] Halevy A, Rajaraman A, Ordille J. Data integration: the teenage years. In: Proceedings of the 32nd international conference on Very Large Data Bases (VLDB), p. 9–16. VLDB Endowment; 2006.
[8] Kopcke H, Rahm E. Frameworks for entity matching: A comparison. Data Knowl Eng 2010;69(2):197–210. McGregor JJ. Backtrack search algorithms and the maximal common subgraph problem. Softw Pract Exp 1982; 12:23–34.
[9] Ngonga Ngomo AC, Auer S. LIMES: a time-efficient approach for large-scale link discovery on the web of data. In: Proceedings of the 22nd international joint conference on Artificial Intelligence. AAAI Press; 2011. p. 2312–2317.
[10] Do HH, Rahm E. COMA—a system for flexible combination of schema matching approaches. In: Proceedings of the 28th international conference on Very Large Databases (VLDB); 2002. p. 610–621.

[11] Peukert E, Eberius J, Rahm E. A self-configuring schema matching system. In: Proceedings of the International Conference on Data Engineering (ICDE); 2012. p. 306–317.
[12] Elmagarmid A, Ipeirotis P, Verykios V. Duplicate Record Detection: A Survey. IEEE T Knowl Data Eng 2007;19(1):1–16.
[13] Thor A, Rahm E. MOMA—A mapping-based object matching system. In: CIDR; 2007.

CHAPTER 3

LEAP Semantics

S. El Kadiri*, A. Milicic*, K. Pardalis and E. Peukert†**
*EPFL, Lausanne, Switzerland
**UBITECH LTD, Athens, Greece
†University of Leipzig, Leipzig, Germany

Contents

3.1 INTRODUCTION

In today's world of fast manufacturing, high quality demands and highly competitive markets, it has become vital for companies to be able to extract knowledge from their operational data, and to manage and to reuse this knowledge in efficient and automated manner. This is particularly relevant in knowledge intensive processes such as product innovation. Furthermore, it's becoming increasingly important to enable the exchange of knowledge between all stages of product life cycle as a prerequisite for optimization of all critical procedures in the design and manufacturing of a product. Ontologies have proven to be of relevance in fulfilling this demand in terms of knowledge acquisition and exploitation, and it has been applied in a number of scenarios within several fields, including medicine, biology, geography, astronomy, defense, and the aerospace industries.

The most appealing features of ontology are: (1) the well-defined structure of the knowledge organization, which is both human and machine understandable; (2) the machine understandable structure enables automatic

Taking the LEAP
http://dx.doi.org/10.1016/B978-0-12-805263-1.00003-7

reasoning and inference, which leads to automatic generation of new knowledge; (3) well-defined semantics enables easy interoperability and design of the plug-in application specific modules; and finally (4) leveraging external ontological resources is enabled through manual or automatic mapping and merging processes, thus eliminating the need for repetition of the entire design process for every application domain.

All of these benefits are vital in the case of large production companies, which are struggling with ever increasing amounts of data and facing problems of exchanging knowledge between different and independently designed systems employed for different stages of the product lifecycle. However, one key downfall of ontology is that it usually has to be manually designed from the beginning for each new use-case. This requires highly specialized knowledge experts working closely with the domain experts for, sometimes, significant periods of time.

The application of ontologies, on the other hand, has grown with the advent of the Linked Data paradigm, and it has become a key enabler for data integration. Linked Data consists of the creation of data stores using URIs to identify resources and their relations, which are annotated using semantics from ontologies. The Linked Open Vocabularies (http://lov.okfn.org/dataset/lov/) initiative, for instance, gives access to an ecosystem of linked open ontologies that are used in the Linked Data Cloud (http://linkeddata.org/). Within this challenging environment, ontology has quickly found its place in academic as well as industrial research and development, leading us to consider its usage as a very promising approach.

3.2 SEMANTIC TECHNOLOGIES

The logical data structure of a database management system (DBMS), either hierarchical, network, or relational, cannot totally satisfy the requirements for a conceptual definition of data because it is limited in scope and biased toward the implementation strategy employed by the DBMS. Therefore, the need to define data from a conceptual point of view has led to the development of semantic data modeling techniques. That is, techniques to define the meaning of data within the context of its interrelationships with other data. The real world, in terms of resources, ideas, events, and so on, is symbolically defined within physical data stores. A semantic data model is an abstraction which defines how the stored symbols relate to the real world. Thus, the model must be a true representation of the real world. The overall goal of semantic data models is to capture more meaning of the data.

The benefits of exploiting semantic data models for business applications are mainly:

1. *Avoid misunderstanding*—by providing a clear, accessible and agreed set of terms and relations as a trusted source, and discussions and misunderstandings can easily be resolved.
2. *Conduct reasoning*—because they are machine understandable, and through the usage of logic statements (rules), ontologies enable automatic reasoning and inference, which leads to the automatic generation of new and implicit knowledge.
3. *Leverage resources*—by extending and relating an application ontology to external ontological resources, via manual or automatic mapping and merging processes, the need for repetition of entire design process for every application domain is eliminated.
4. *Improve interoperability*—semantic models can serve as a basis for schema matching to support systems interoperability in close environments where systems, tools, and data sources have no common recognition of data type and relationships.

Ontologies play an important role for many knowledge-intensive applications because they provide formal models of domain knowledge that can be exploited in different ways. Ontology development has become an engineering discipline, Ontology Engineering, which refers to the set of activities that concern the ontology development process and the ontology life cycle, the methods and methodologies for building ontologies, and the tool suites and languages that support them [1].

3.2.1 Ontology

3.2.1.1 Definitions, Objectives, and Benefits

The word ontology was taken from philosophy, where it means a systematic explanation of being. There are many definitions of ontology, which have changed and evolved over the years. However, [2] provide one of the most well-known definitions: "An ontology is a formal, explicit specification of a shared conceptualization. Conceptualization refers to an abstract model of some phenomenon in the world by having identified the relevant concepts of that phenomenon. "Explicit" means that the type of concepts used, and the constraints on their use are explicitly defined. "Formal" refers to the fact that the ontology should be machine-readable. "Shared" reflects the notion that an ontology captures consensual knowledge, that is, it is not private of some individual, but accepted by a group."

Ontologies can be modeled with different knowledge modeling techniques and they can be implemented in various kinds of languages based on different knowledge representation formalisms. It is important to mention here that there are many connections and implications between the knowledge modeling components (concepts, roles, etc.) used to build an ontology, the knowledge representation paradigms (frames, description logics, logic) used to represent formally such components, and the languages used to implement the ontologies under a given knowledge representation paradigm. However, they all share the following minimal set of components:

- *Classes* represent concepts, which are taken in a broad sense. For instance, in the domain of the Energy Efficiency of Buildings, the concepts are Building, Door, Window, Device, Sensor, and so on. The classes in the ontology are usually organized in taxonomies, through which inheritance mechanisms can be applied. We can represent a taxonomy of sensors (Scanning Sensor, Optical Sensor, Touch Trigger Sensor, etc.) or different types of doors in the buildings (Inner Door, Outer Door, Sliding Door, Rotating Door, or Strong room Door).
- *Relations* represent a type of association between concepts of the domain. They are formally defined as any subset of a product of *n* sets, that is: $R \subset C1 \times C2 \times \ldots \times Cn$. Ontologies usually contain binary relations. The first argument is known as the domain of the relation, and the second argument is the range. For instance, the binary relation locatedIn has the concept Building as its domain and the concept Location as its range. In addition, this relation can have the concept Device as a domain. Binary relations are sometimes used to express concept attributes (also known as slots). Attributes (name, version, weight, etc.) are usually distinguished from relations (isComposeOf, hasConstraint, hasParameters, etc.) because their range is a datatype, such as string, number, and so on, while the range of relations is a concept.
- *Formal axioms*, according to [1], formal axioms serve to model sentences that are always true. They are normally used to represent knowledge that cannot be formally defined by the other components. In addition, formal axioms are used to verify the consistency of the ontology itself or the consistency of the knowledge stored in a knowledgebase. Formal axioms are very useful to infer new knowledge. An axiom in the Energy Efficiency of Buildings domain could be that it is not possible to build a public building without a fire door (based on legal issues).
- *Ontology rules* have been the subject of much debate over whether they should be adopted as a part of the ontology structure or as a layer on

top of the ontology. Despite this debate, ontology rules are a mechanism to expand the functionality of the ontology by modeling the logic and dynamics of the domain. More detail about ontology rules will be given in Section 3.3.3.

- *Instances* are used to represent elements or individuals in an ontology.

3.2.1.2 Development of Ontology

Until 2009, METHONTOLOGY (http://semanticweb.org/wiki/METHONTOLOGY), On-To-Knowledge (http://www.ontotext.com/research/otk), and DILIGENT (http://semanticweb.org/wiki/DILIGENT) were the most common methodologies for building ontologies. These methodologies mainly include guidelines for single ontology construction, ranging from ontology specification to ontology implementation, and they are mainly targeted at ontology researchers. In contrast, a new methodology, called the NeOn Methodology (http://www.neon-project.org/), suggests pathways and activities for a variety of scenarios, instead of prescribing a rigid workflow.

The *NeOn Methodology* [3] for building ontology networks is a scenario-based methodology that supports a knowledge reuse approach, as well as collaborative aspects of ontology development and dynamic evolution of ontology networks in distributed environments. A network of ontologies is a collection of ontologies related together via a variety of relationships, such as alignment, modularization, version and dependency. The key assets of the NeOn Methodology are:

- A set of nine scenarios for building ontologies and ontology networks, emphasizing the reuse of ontological and nonontological resources, reengineering and merging, and taking collaboration and dynamism into account.
- The NeOn Glossary of Processes and Activities, which identifies and defines the processes and activities carried out when ontology networks are collaboratively built by teams.
- Methodological guidelines for different processes and activities of the ontology network development process, such as the reuse and reengineering of ontological and nonontological resources, ontology requirements specification, ontology localization, scheduling, and so on. All of the processes and activities are described with (1) a filling card, (2) a workflow, and (3) examples.

METHONTOLOGY [4] enables the construction of ontologies at the knowledge level. It includes (1) the identification of the Ontology

Development Process (ODP) that defines which tasks should be performed when building ontologies; (2) a life cycle that is based on evolving prototypes; and (3) some techniques to carry out management, development-oriented, and support activities. In addition, METHONTOLOGY includes a list of activities to be carried out during ontology reuse and reengineering processes but it does not provide detailed guidelines for such activities nor does it consider different levels of granularity during the reuse of ontological resources (eg, modules or statements). Moreover, METHONTOLOGY does not consider the reuse and reengineering of nonontological resources or the reuse of ODP.

On-To-Knowledge [5] proposes that ontologies should be built by taking how they are going to be used in knowledge management applications into account. The processes proposed by this methodology include the following: feasibility study; kickoff, where ontology requirements are identified; refinement, where a mature and application-oriented ontology is produced; evaluation; and maintenance. With respect to the reuse of knowledge resources, the kickoff process mentions that developers should look for potentially reusable ontologies. However, this methodology does not provide detailed guidelines to identify such ontologies or tell us how to reuse them. Besides, this methodology does not explicitly mention guidelines for the reuse and reengineering of nonontological resources, or for the reuse of ontology design patterns.

The *DILIGENT* methodology [6] is intended to support domain experts in a distributed setting in order to engineer and evolve ontologies. This methodology is focused on collaborative and distributed ontology engineering. Its ontology development process includes the following five activities: building, local adaptation, analysis, revision, and local update. With regard to the reuse of knowledge resources, this methodology does not include guidelines for the reuse and reengineering of existing knowledge resources.

3.2.2 Ontology-Based Reasoning

The domain knowledge often contains dynamics between concepts and processes which are triggered only when certain set of conditions are met.

Ontology rules when used as a component of ontology structure enable us to model the dynamics of the domain, functional, and interdependent relations between concepts and their properties. In general, ontologies concentrate on classification methods, which places emphasis on: defining 'classes' and 'subclasses'; how individual resources can be associated to such classes; and characterizing the relationships among classes and their instances. Rules, on the other hand, concentrate on defining a general mechanism on discovering and generating new relationships based on existing ones, much like logic programs. In the ontology design phase, an initial set of rules is manually defined, so that it implements logic, conditional processes, and inherence structure. It is then left to inference engines (or reasoners) to automatically generate every valid conclusion. Several approaches are used for inference engines:

1. *Description logic-based inference engine*—This is used to perform basic reasoning tasks such as consistency checking and subsumption concepts. It has the advantage of using decidability
2. *First order logic* (*FOL*)—The reasoner takes an OWL file as an input and first translates this into FOL. The inference is then processed by using any one of the existing automated theorem provers.
3. A combination of FOL and general logic-based inference engine—In this approach, a fragment of FOL and general logic is used to design an OWL inference engine. Horn logic is most widely used due to its simplicity and availability

We have chosen to use the Pellet reasoning engine. As with most of the inference engines, Pellet works on a knowledge base composed out of TBox (semantic layer where concept ontology is defined) and ABox (where instances are defined). The reasoning services that are applied on this knowledge base are:

1. *Consistency Checking*—Consistency checks whether ABox instances are consistent with the TBox concepts and ensures that ABox meets all of the restrictions.
2. *Satisfiability Checking*—This checks if OWL concepts cover instances according to the current ontology. This means that ABox instances should be satisfied by the available OWL concepts of TBox.
3. *Subsumption Checking*—This checks whether a class D subsumes another class C. This means that instances or properties of class C are also part of class D.
4. *Query Processing*—OWL inference engines need a powerful language to support queries so that users and software agents can query a knowledge base to retrieve useful data or facts.
5. *Reasoning with Rules*—Because rules are capable of expressing OWL classes, properties, and instances, the OWL inference engine needs to provide an interface to process rules that are represented with OWL classes, properties, and instance data.

3.2.3 Ontology Querying

An ontology as an overall model of a domain and the knowledge in it can be used as a research tool for all users. If the user is new to a domain and wants to analyze the relationships between certain objects, he or she might not have an exact model in mind. One approach to this problem is to apply data mining and identify interesting combinations of dimensions or to find frequent item sets. Another approach is recommendation. If one user as already asked for a specific pattern in the graph data, then this pattern could also be interesting to other users, even though only parts of the pattern are reusable. Moreover, if the domains are similar enough, one could translate interesting pattern queries from one domain to another and get new recommendations.

To explain this idea, consider a simple example of a software development process and a knowledge engineering process. Both work on artifacts which are code fragments in software development and knowledge documents in the other domain. Both domains have the quality criteria of the artifacts, which can be lines of code, number of errors in the software development, or the ratings or number of references and reads in the other domain. A query that looks for the developers that have written code with many errors could be translated to a query in the knowledge management domain that searches for people who write knowledge documents with a bad rating.

Fig. 3.1 shows a screen shot of a rather technical prototype of the query recommendation functionalities that we have developed. An input pattern query given in the top-left contains a number of types and edges. In the example, only one type and one edge is contained. On the right hand side, we list a number of currently stored queries by name. A user could now store a query in the ontology or ask for recommendations. The recommendations are listed on the bottom. Here, the recommendation engine gives a query that also involves a persona type that did not exist in the initial query. A user could take that query and copy, modify, and execute it. Modified queries can then also be stored again so that the query store can grow over time.

Luckily, a lot of effort was put into the design of the LEAP ontology that encodes key concepts from three different domains and which defines the abstract concepts that all of the domains inherit from. We developed a recommendation engine that allows us to translate pattern queries to statements in the ontology, which can be stored. Moreover, based on the inheritance relations, these statements can be translated to other domains with the use of SPARQL queries. The resulting derived statements are finally

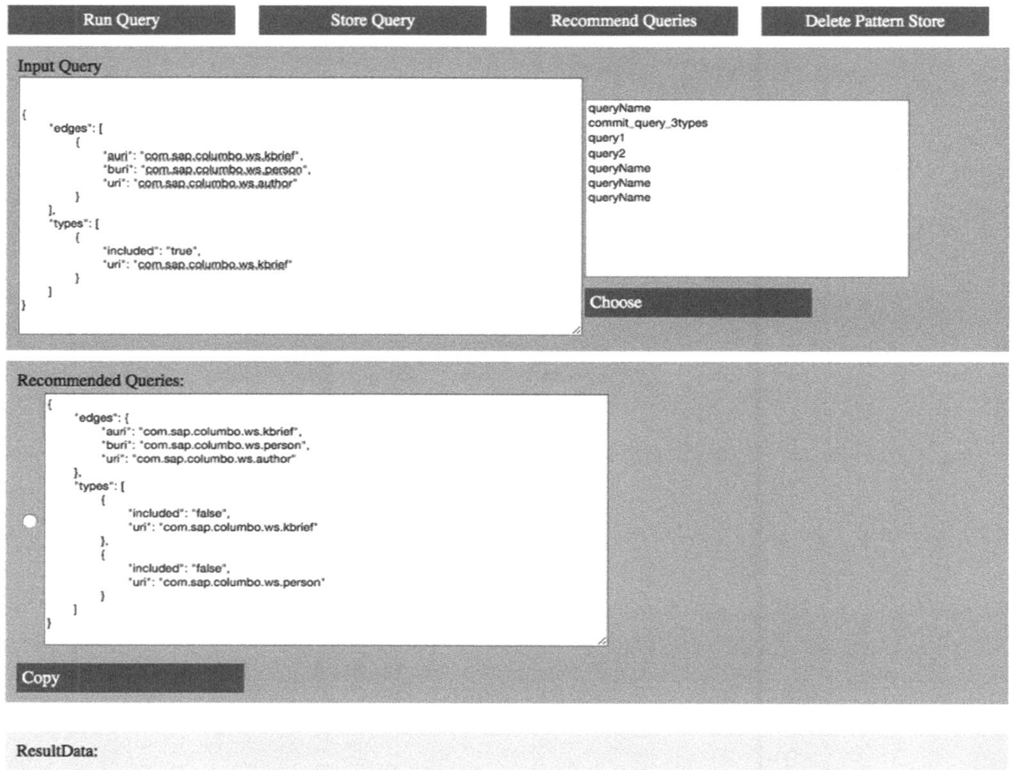

Figure 3.1 *Query recommendation prototype.*

translated back to pattern queries, which can then be executed or presented to the user as recommendations.

3.3 LEAP ONTOLOGY AND RULE BASE

3.3.1 Requirements for the Semantic Modeling of LEAP Use Cases

Quality control in the *Volkswagen* and TRIMEK use-cases refers to the hot stamping process and aims at analysing correlations between production process parameters, geometric and microstructural properties, and geometric deviations revealed within work pieces, thus enabling an even better quality, helping to reduce errors and defects, and finally improving and adapting the production process. In order to automatically detect pieces with defects, TRIMEK provided a 3D method to scan the manufactured piece and to generate its CAD representation. Errors are than detected by comparing the CAD generated with the TRIMEK system and that of the ideal Master piece.

A relevant problem has to be noticed in this use case, a scanning process is time consuming, leading to the conclusion that it is not feasible to scan each piece produced. The proposed solution is to monitor different conditions during the process, such as temperature, pressure, and time to detect when some of these parameters might be a cause of geometric deviations and then declare that piece as "risky". The piece in question then will be scanned to determine whether or not there is a defect. This means that scanning is eliminated as a bottle-neck in the production procedure because only a certain subset of pieces will be scanned. The solution relies on an assumption that the cause of the defects lies within the monitored parameters. To embed these heuristics into an ontological knowledge base, we have interviewed the workers and modeled what they treat as critical parameter values into ontological rules.

Given that the ontology is designed so that it covers the entire procedure of quality control, ontology rules are defined to support the scanning results, also capturing appropriate actions for when the defect is detected.

The COMAU use-case involves mostly two different phases of the product lifecycle: the proposal (or concept) phase and the operational phase. The goal of this use-case is to have an automated system to calculate the LCC of COMAU products (in this case, an engine assembly line) and to optimize the line configuration according to the costs and environmental impact. This is done in the proposal phase in such a way that more information can be attached to the offer that COMAU presents to its customers. However, the data used to calculate the LCC comes from databases that can be refined by collecting information from the shop-floor where the equipment is working. This connection between the operational phase and the concept phase of the COMAU equipment is reached through an automated data collection system that is able to record all of the relevant information coming from the machine's control system (production data, failure data, maintenance activities, energy consumption, etc.). In the proposal/concept phase, the designers take decisions on the line configuration to be presented to the customer. Assembly lines are generally made of many stations, where a certain set of operations are performed either by an operator or by automatic equipment. A module was developed within LEAP that automatically finds the configuration with the lowest LCC value; however, it is not able to take into consideration the more complex requests from the user. A list of potential requests is given as a list of properties of the concept 'Customer' in the COMAU ontology.

In the *Aker* use-case, knowledge is stored and structured using the K-brief methodology. Employees capture their experience and work in digital forms, which is then exchanged, browsed, and retrieved. To be able to maintain a high quality knowledge base, K-briefs need to be reviewed and validated. For this purpose, we introduced a property of every K-brief, named 'Maturity'. The first requirement for the knowledge management ontology in this case is to be able to maintain the value of maturity for every K-brief. Second, to enable easy retrieval of knowledge the K-briefs are tagged with keywords according to their content, and the tags need to be manually added once the K-brief is created. Considering that the ontology contains the relations between a product, its parts, and modules, a second requirement is to enable automatic recommendation of additional keywords, based on a set of existing ones. Furthermore, since Aker carries out projects in different parts of the world, two identical oil-drilling platforms will have to follow different industrial standards depending on their location. For this reason, we use the ontology to enable automatic industrial standard recommendation, based on the project's characteristics. Finally, within LEAP, we propose a novel approach for industrial standards application. The idea is to model the standard into ontology, adding constraints in a form of concept properties and we then use the rules to control the designer decisions. Considering that the number of standards used is very high, only a methodology and example of its application are provided.

In summary, the requirements cover aspects from three main product lifecycle stages:

1. *Product definition stage*—In this stage, the main requirement is to enable lifecycle cost analysis and assessment in order to optimize engineering

decisions and facilitate efficient allocation of resources. For more details about this case please see Section 8.2. The Lifecycle Cost Assessment Ontology is proposed to cover this requirement and it can be reused in different industrial contexts.

2. *Product design stage*—In this stage, the main requirement is to support knowledge capture and reuse. For more details about this case please see Section 8.3. The Design Knowledge Ontology is proposed to cover this requirement and it can be reused in different industrial contexts.
3. *Product manufacturing stage*—In this stage, the main requirement is to support the reduction of errors and defects in part/product manufacturing in order to improve and adapt the production process. For more details about this case please see Section 8.1. The Quality Control Ontology is proposed to cover this requirement and it can be reused in different industrial contexts.

3.3.2 LEAP Ontology Network

The LEAP ontology network consists of an upper ontology and its specialization to three domain specific ontologies related to the industrial use-cases: Volkswagen-TRIMEK, COMAU, and Aker Solutions.

The main reason to have an upper ontology and domain specific ontologies in a top-down approach is to establish a well-defined meaning (semantics) that is consistent across contexts, and which can be easily adjusted and adopted for different application domains. The other reason is that the three LEAP use-cases deal with different scenarios (and needs) but have a common objective that is knowledge elicitation and exploitation for the design and production of future products.

The upper ontology is a core metamodel representing a minimum and generic set of concepts and relationships, and which can be considered domain-independent. A domain specific ontology represents a domain-specific description that addresses its particular set of requirements, and should also have the minimum set of concepts and relationships required for describing its determined domain. The addition of domain specific ontologies should have a minimum impact on the upper ontology.

3.3.2.1 Upper Ontology

This subsection provides a list of the concepts and object properties of the upper ontology. The graph representation of the upper ontology is described in Fig. 3.2. For visibility reasons only parent–child relationships are displayed in this graph.

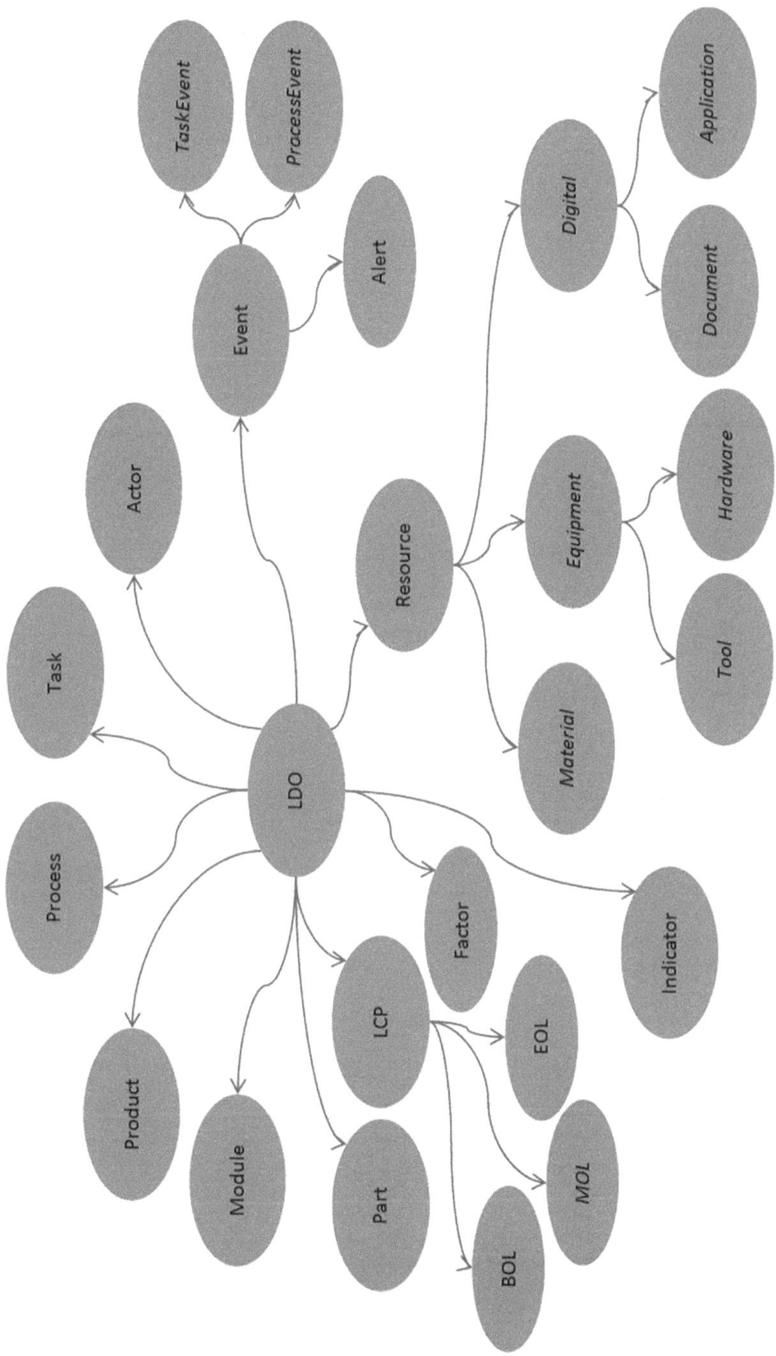

Figure 3.2 ***Graph representation of upper ontology.***

Table 3.1 Upper ontology concepts and subconcepts

High level concepts	Subconcepts	Subconcepts	Description
Task	—	—	Groups scheduled actions
Process	—	—	Groups all processes
Product	—	—	Groups all products
Module	—	—	Groups modules within a product structure
Part	—	—	Groups parts within a product module
Actor	—	—	Groups all the persons involved in an organization
Factor	—	—	Groups relevant factors related to evaluate a product or process
Indicator	—	—	Groups metrics applied to evaluate a factor
Resource	Digital	Document Application	Represents digital resources in terms of Documents and Applications
	Equipment	Tool Hardware	Equipment in terms of tools and hardware used to perform a specific task
	Material		Represents raw material needed to produce a product
LCP	BOL		Beginning Of the Life Cycle
	MOL		Middle Of the Life Cycle
	EOL		End of Life Cycle
Event	TaskEvent		Triggers a task
	ProcessEvent		Triggers a process
Alert			Notifies relevant actors

A detailed list of the upper ontology concepts, their subconcepts, and their description is provided in Table 3.1.

Table 3.2 lists the object properties of the upper ontology. The domain and range are defined for each property.

Table 3.2 Object properties of upper ontology

Relation	Domain	Range
generatesTask	Event	Task
stimulatesActor	Event	Actor
stimulatesApplication	Event	Application
triggersAlertNotification	Event	Alert
triggersTask	TaskEvent	Task
triggersProcess	ProcessEvent	Process
generatesEvent	Process	Event
belongsTo	Process	LCP
hasTask	Process	Task
associatedWith	Process	Product
comprisedOf	Product	Module
hasModule	Module	Module
comprisedOf	Module	Part
associatedWith	Task	Part
uses	Task	Resource
hasFactor	Task	Factor
hasIndicator	Factor	Indicator
decidesUpon	Actor	Event
triggersEvent	Application	Event
performs	Actor	Task

3.3.2.2 Quality Control

This subsection provides the list of concepts and object properties of the quality control ontology. The graph representation of the ontology is described in Fig. 3.3. For visibility reasons, only the parent–child relationships are displayed in this graph.

The detailed list of the upper ontology concepts, their subconcepts, and their description is provided in Table 3.3.

Table 3.4 lists the object properties of the quality control ontology. The domain and range are defined for each property.

3.3.2.3 Life Cycle Cost Assessment

This subsection provides the list of concepts and objects properties of the Life Cycle Cost assessment (LCC) Ontology. The graph representation of the ontology is described in Fig. 3.4. For visibility reasons, only the parent–child relationships are displayed in this graph.

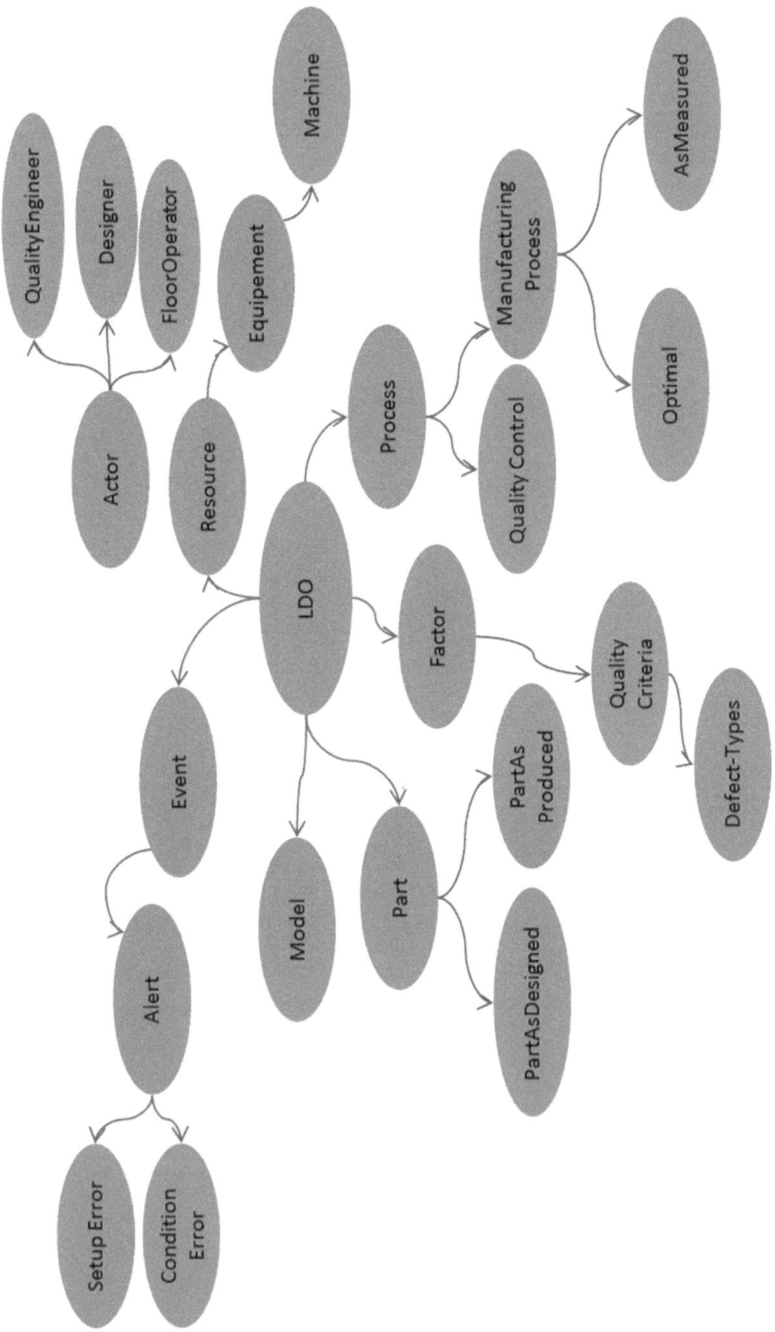

Figure 3.3 *Graph representation of quality control ontology.*

Table 3.3 Quality control ontology concepts and description

High level concepts	Description
MaufacturingProcess	Defines the production line's conditions and generally represents sequential stages of the manufacturing process including different tools and processes
Optimal	Optimal process is the concept defining the optimal parameters of the production line
As Measured	This concept models the process as actually performed
Machine	Models the machines and tools used in each production step. One of the attributes describing it is Maintenance Period
PartAsDesigned	Defines the master piece represented by the CAD model of a part which is considered to be ideal as designed
PartAsProduced	Defines the produced part and can be represented by a 3D model generated from 3D scanning
QualityCriteria	Groups aspects considered in quality control. On aspect could be all known aberrations that can occur during production and for which the part is tested
SetupErrors	Signals to relevant actors the problem in settings of a manufacturing line
ConditionErrors	Signals to relevant actors problems about the manufacturing process
Defect-Types	Defect types are undesirable deviations of planned characteristics of a product
Designer	Designer is a person in charge of product design
FloorOperator	Floor operator is a worker in charge of monitoring the manufacturing process
QualityEngineer	Quality engineer is a person in charge of the aberration detection process as well as reactions to a defect confirmation

Table 3.4 Object properties of quality control ontology

Relation	Domain	Range
Described by	Part	PartAsDesigned
Described by	Part	PartAsProduced
Considers	PartAsProduced	QualityCriteria
IsPrefered	ManufacturingProcess	Optimal
IsPerformed	ManufacturingProcess	AsMeasured
Uses	AsMeasured	Machine

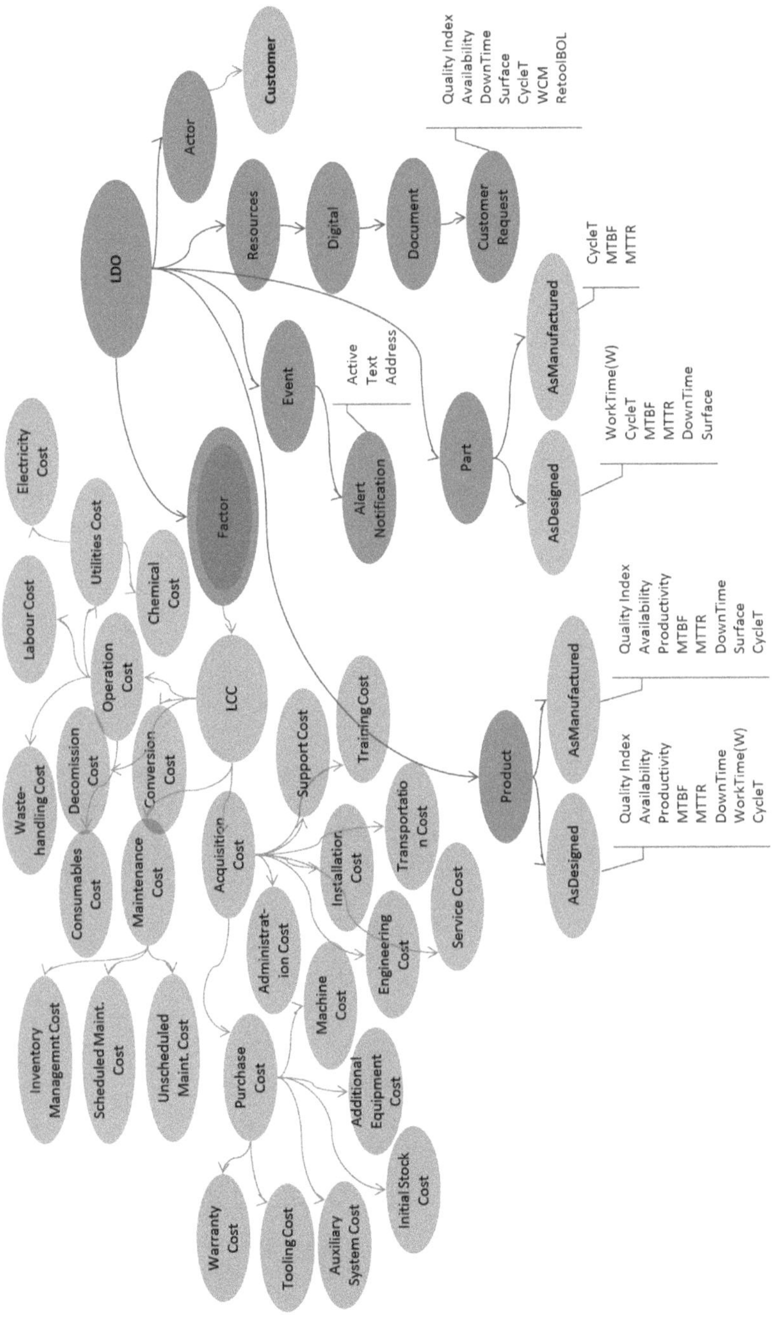

Figure 3.4 *Graph representation of LCC ontology.*

All of the concepts of LCC ontology and their description are listed in Table 3.5; the hierarchy Concept/Subconcept is given in Table 3.6, Table 3.7, Table 3.8, and Table 3.9.

Table 3.5 LCA ontology concepts and subconcepts

High level concepts	Subconcepts	Description
LCC	Acquisition Cost	It includes Concept, Development & Design and Built & Install stages of the life cycle phases so it represents the money required to have the machine in full operating mode in the plant. It can be calculated before the acquisition is made and it is the tip of the lifecycle cost iceberg. The acquisition cost is unique in the sense that you only have to pay once for it.
	Operation Cost	This is the money the installation costs when it is running. It doesn't include the moments when the machine is down or stopped but only when it is operational. Unlike the acquisition cost, the operation cost is recurring.
	Maintenance Cost	This cost represents the money and the work required to keep the asset in proper running condition. It also means that the machine is not operational during this time and that's why this cost is independent from the operation cost. The maintenance can be scheduled (eg, for the general overhaul) or not (because of an unexpected failure). This is a recurring cost.
	Conversion Cost	This cost is necessary when the machine needs to be converted to handle the production of different components than the ones it was initially designed for. It could become a recurring cost if many conversions of the machine are made during its life.
	Decommission Cost	The decommission cost is the money required when you want to get rid of the machine. To do so, you can either dismantle totally the installation or you can sell it (or at least some of the components). It is worth noting that in the latter case, it would be a revenue and not a cost. This should be taken into account for the LCC analysis, even if it is a tough estimation to make.

(Continued)

Table 3.5 LCA ontology concepts and subconcepts (*cont.*)

High level concepts	Subconcepts	Description
Indicator	MTBF	MTBF is the predicted elapsed time between inherent failures of a system during operation. MTBF can be calculated as the arithmetic mean (average) time between failures of a system.
	MTTR	MTTR is a basic measure of the maintainability of repairable items. It represents the average time required to repair a failed component or device.
	Availability	Availability is the proportion of time a system is in a functioning condition.
	Cycle Time	Cycle time is the total time from the beginning to the end of a process
	OEE	OEE is a hierarchy of metrics to evaluate how effectively a manufacturing operation is utilized.
	UP Time	Uptime is a measure of the time a machine has been working and available. Uptime is the opposite of downtime.
	Down Time	The term Down Time is used to refer to periods when a system is unavailable. Downtime or outage duration refers to a period of time that a system fails to provide or perform its primary function.
	Floor space required	The floor space required to accommodate the equipment.

LCC, life cycle cost; *MTBF*, mean time between failures; *MTTR*, mean time to repair; *OEE*, overall equipment effectiveness.

Table 3.10 lists the object properties of the LCC ontology. The domain and range are defined for each property.

3.3.2.4 Knowledge-Briefs in Automated Design

This subsection provides the list of concepts and objects properties of the design knowledge ontology. The graph representation of the ontology is described in Fig. 3.5. For visibility reasons, only the parent–child relationships are displayed in this graph.

All of the concepts of Design Knowledge Ontology, their subconcepts, and their description are listed in Table 3.11.

Table 3.12 lists the object properties of the LCC ontology. The domain and range are defined for each property.

Table 3.6 LCC concepts and subconcepts

LCC concepts	Subconcepts	Description
Acquisition Cost	Administration Cost	Cost associated to the administration (travels, …)
	Engineering Cost	Cost of all the engineering required to develop a new machine
	Installation Cost	Cost to install the machine in the customer's plant
	Packaging Cost	Cost of the packaging required to transport the machine
	Purchase Cost	Cost of the different items and equipment required to have a fully operative installation
	Service Cost	Cost to provide a 24-h service to the customer
	Support Cost	Cost to provide an efficient support to the customer (upgrades, updates, …)
	Training Cost	Cost to train the operators for the new machine
	Transportation Cost	Cost to transport the asset from where it has been built to the factory where it'll be used
Operation Cost	Consumables Cost	Cost of all the consumable items used
	Labor Cost	Cost of the operator (maintenance operators, M/C operators, …)
	Utilities Cost	Cost of the total utilities consumption (chemicals and electricity)
	Waste Handling Cost	Cost related to collecting and disposing of waste products
Maintenance Cost	Inventory Management Cost	Cost to manage the inventory and to bring it back to its initial values (wear parts, spare parts, and standby parts) after a maintenance
	Scheduled Maintenance Cost	Cost required to perform all the different scheduled forms of maintenance
	Unscheduled Maintenance Cost	Cost of the unplanned maintenance, which means the machine was supposed to be producing during these times

Table 3.7 Acquisition cost hierarchy

Acquisition cost hierarchy	Subconcepts	Subconcepts
Administration Cost	Travel Cost	—
Engineering Cost	Development Cost	—
	Runoff Cost	—
	Try out Cost	—
Installation Cost	—	—
Packaging Cost	—	—
Purchase Cost	Additional Equipment Cost	—
	Auxiliary Systems Cost	Coolant Filtration System Cost
		Hydraulic System Cost
		M/C PLC Cost
	Initial Stock Cost	Initial Spare Parts Cost
		Initial Standby Parts Cost
		Initial Wear Parts Cost
	Machine Cost	—
	Tooling Cost	Clamping Unit Cost
		Setting Gauge For Tooling Cost
		Tool Holder Cost
		Tools Cost
	Warranty Cost	Additional Warranty Cost
Service Cost	24hService Cost	—
	24hService for M/C Cost	—
Support Cost	Hardware Support Cost	—
	Software Support Cost	—
Training Cost	—	—
Transportation Cost	—	—

Table 3.8 Maintenance cost hierarchy

Maintenance cost hierarchy	Subconcepts
Inventory Management Cost	Spare Parts Stock Per Year Cost
	Standby Parts Stock Per Year Cost
	Wear Parts Stock Per Year Cost
Scheduled Maintenance Cost	General Overhaul Labour Per Year Cost
	General Overhaul Material Per Year Cost
	M/C-operator Labour Per Year Cost
	Predictive Maintenance Per Year Cost
	Preventive Maintenance Per Year Cost
Unscheduled Maintenance Cost	Corrective Maintenance Labour Per Year Cost
	Unplanned Downtime Per Year Cost

Table 3.9 Operation cost hierarchy

Operation cost hierarchy	Subconcepts	Subconcepts
Consumables Cost	Coolant Per Year Cost	—
	Detergent Per Year Cost	—
	Hydraulic Oil Per Year Cost	—
	Lubricant Per Year Cost	—
	Miscellany Consumption Per Year Cost	—
	Quench Oil Per Year Cost	—
	Toolset Per Year Cost	—
Labour Cost	M/C-operator Labour Cost Per Hour	—
	Maintenance Labour Cost Per Hour	—
	Tool Change Labour Per Year	—
Utilities Cost	Decommission Cost	—
	Chemicals Cost	Ammonia Per Year Cost
		Compressed Air Per Year Cost
		Demineralized Water Per Year Cost
		Effluents Per Year Cost
		Industrial Water Per Year Cost
		Inert Gas Per Year Cost
		Methanol Per Year Cost
		Natural Gas Per Year Cost
		Nitrogen Per Year Cost
		Other Gas Per Year Cost
		Oxygen Per Year Cost
		Propane Per Year Cost
	Electricity Cost	Electricity In Cycle Per Year Cost
		Electricity In Idle Per Year Cost
Waste Handling Cost	Waste Disposal Per Year Cost	—
	Waste Handling Per Year Cost	—

3.3.3 Rule Base Set

As explained in Section 3.2.2, the knowledge of dynamics in the domain can be modeled by ontology rules, so that it creates and adds values of the ontology functionality. As will be shown, ontology can act as an alert system, design consistency control, knowledge maintenance system, and many more.

Table 3.10 Object properties of LCA ontology

Relation	Domain	Range
requiresPackaging	Transportation Cost	PackagingCost
dependsOn	GeneralOverhaulLabour PerYearCost	MaintenanceLabour CostPerHour
	M/COperatorLabour PerYearCost	M/COperatorLabour CostPerHour
	CorrectiveMaintenance LabourPerYearCost	MaintenanceLabour CostPerHour
rnvolves	ScheduledMaintenanceCost	InventoryManagementCost
	UnscheduledMaintenance-Cost	
reduces	PredictiveMaintenance PerYearCost	UnscheduledMaintenance Cost
	PreventiveMaintenance PerYearCost	

The Semantic Web Rule Language (SWRL) is a proposed language for the Semantic Web that can be used to express rules as well as logic, combining OWL DL or OWL Lite with a subset of the Rule Markup Language (itself a subset of Datalog). SWRL has the full power of OWL DL, but at the price of decidability and practical implementations. However, decidability can be regained by restricting the form of admissible rules, typically by imposing a suitable safety condition. Rules are of the form of an implication between an antecedent (body) and consequent (head). The intended meaning can be read as: whenever the conditions specified in the antecedent hold, then the conditions specified in the consequent must also hold. For example, in the VOLKSWAGEN case, the rules contain knowledge related to the hot stamping process and how a number of specific measured parameters can lead us to the conclusion that errors might appear to the resulting product. These errors take the form of geometric deviations and overlaps. An example of such a rule can be found in Table 3.13.

This rule refers to concepts that appear in the Quality Control specialization of the LEAP Ontology but it is not yet expressed in a form that can be interpreted by a machine. As mentioned in the previous sections, the rule language SWRL was selected.

Using this language, the rule in Table 3.13 is expressed as presented in Table 3.14.

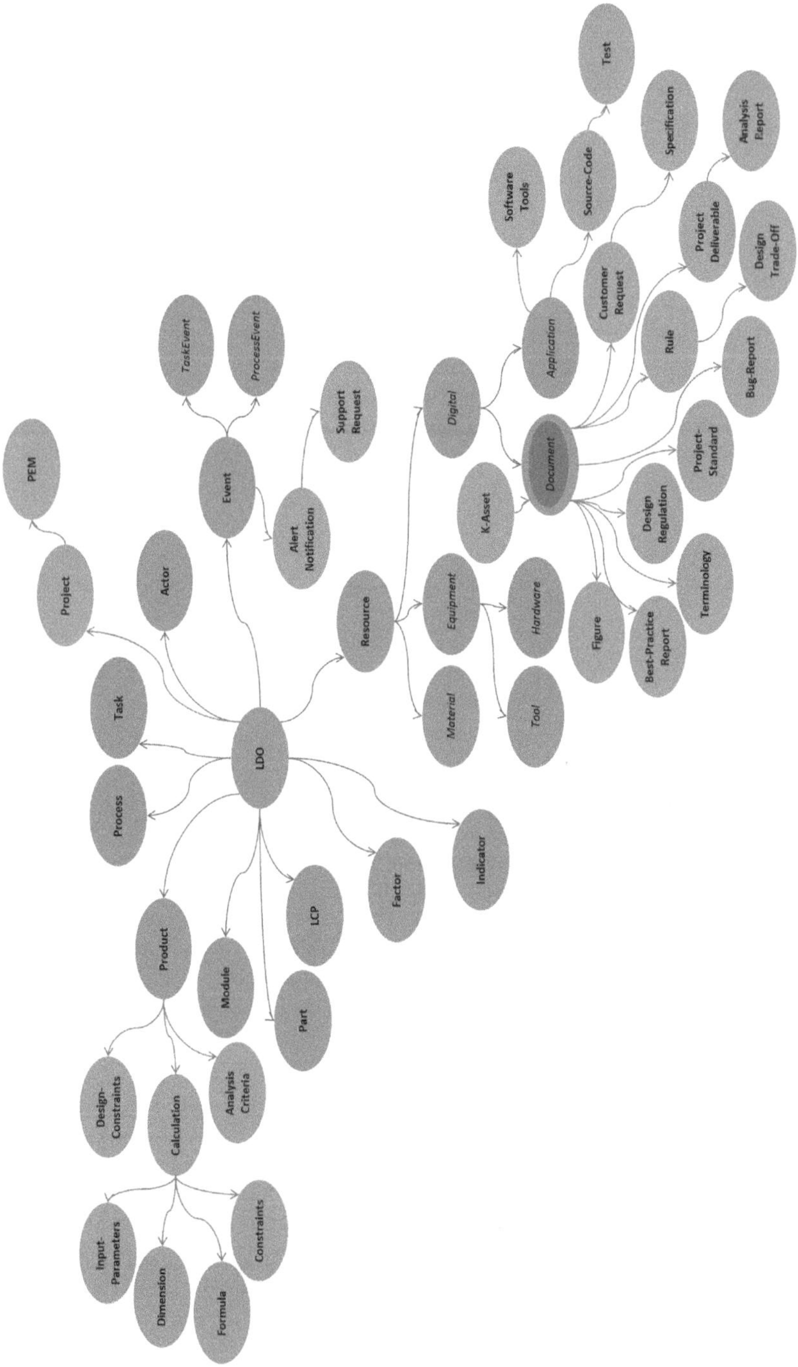

Figure 3.5 *Graph representation of design knowledge ontology.*

Table 3.11 Design knowledge ontology concepts and subconcepts

High level concepts	Subconcepts	Description
Design-Constraint	—	Describes the restrictions of a design and it is used as input for the definition of rules.
Input-Parameter	—	Describes all input parameters needed for Calculations. It has the Author ID as an attribute (among others).
Constraint	—	Describes constraints on Input parameters, Dimensions and Calculation results. The concept is taken from the MOKA Informal model.
Formula	—	Describes a mathematical relationship or rule expressed in symbols. It is used for mathematical calculations.
Dimension	—	Describes things that can be annotated on a drawing and measured. It can be a distance, an angle or a radius on the item studied. Attributes describing this concept are Value, Unit (SI unit). Dimensions can be an input or the result of a calculation.
Calculation	—	Describes the result of the Constraints and Formulas when applied to Input parameters. Calculation has Status as an attribute (among others).
Analysis-Criteria	—	Describes criteria used to analyze the structure of a given Module.
Rule	Design-trade-off	Subconcept of Document. Rule is a human understandable explanation of logic, constrains and principles applied in the project. It has a subconcept Design-trade-off. It shows for instance an established level of trade-off between fabrication cost and optimizing weight.
Terminology		Subconcept of Document, storing the glossary.
Design-regulation		Subconcept of Document, defining the regulation with respect to the design code (eg, API).
Project-deliverable	Analysis-report	Subconcept of Document. It is an output of the engineering department. It can be 3D models, analysis results, drawings, and so on. It has a relation to the concept Project (described hereafter).
Bug-report		Subconcept of Document, storing an error or a problem with the software which is reported to support.
Bestpractice-report		Subconcept of Document. These reports are documents storing lessons learned from projects regarding critical-to-quality and critical-to-cost issues. It is subconcept of Document.

Table 3.11 Design knowledge ontology concepts and subconcepts (*cont.*)

High level concepts	Subconcepts	Description
Project-standard		Subconcept of Document. It groups all the general and industrial standards giving specific directions relevant for a project.
Customer-request	Specification	Subconcept of Document, usually text or graphical file describing requirements in general terms. It has subconcept Specification that containes functional specification of the product in details.
Figure/illustration		Subconcept of Document.
K-Asset	—	Represents a Knowledge Asset. It may include important design trade-offs and decisions, reusable design elements, solutions to critical-to-quality issues, solutions to critical-to-cost issues, performance curves, raw material/component data, test results for common design elements reliability/environmental data, factory design rules/capability data, supplier design rules/capability data frequently used parts/raw materials. It contains a link to other Documents. Attributes describing this document are (among others) the Author ID and Description (as a string attribute), which is used to organize the Brief.
Source-Code		Subconcept of Application concept containing source code representation of a design object (a Module/Part). The user has a possibility to inspect source code of classes that implement a design object.
Software-tools		It represents a set of analysis or design software used. It's subconcept of Application.
Support-request		Represents a request for support. It is a sub concept of Alert noting the requests and alerts from the customer.
PEM		Stands for Project Execution Model. This concept contains standards for execution of different kind of projects. Given a Project type, PEM defines a specific way of proceeding which is specified in the PEM. PEM could be used to track decisions in earlier projects and to see how issues have been solved.
Project		Deals with project such as for delivering early cost estimates, drawing packages, 3d models, analysis models, and so on.
Discipline		Category to which the product design belongs to.

Table 3.12 Object properties of design knowledge ontology

Relation	Domain	Range
isUsed	Analysis-Criteria	Module
usesMaterial	Part	Material
applies	Rules	Design-Constraint
usesParameter	Calculation	Input-Parameter
hasConstraint	Input-Parameter	Constraint
	Dimensions	Constraint
	Calculation	Constraint
usesFormula	Calculation	Formula
belongs	K-Asset	Discipline
associatedTo	Project	PEM
createdBy	K-Asset	Actor

Table 3.13 An example of a rule

If HotStamping.AsMeasured.Heating.DwellTime > HotStamping.Optimal.DwellTime then ConsideredRisky = 1

Table 3.14 Rule given in Table 3.13 expressed in SWRL

As_Measured(?m), Heating(?h), Optimal(?o), hasDwellTime(?h, ?x), hasDwellTime(?o, ?y), greaterThan(?y, ?x) -> ConsideredRisky(?m, false)
As_Measured(?m), Heating(?h), Optimal(?o), hasDwellTime(?h, ?x), hasDwellTime(?o, ?y), greaterThan(?x, ?y) -> ConsideredRisky(?m, true)

The rules base sets developed for each of the LEAP use-cases are presented in the next subsections.

3.3.3.1 Rules for Quality Control

The rule base set for quality control defined for VOLKSWAGEN use-case contains three types of rules:

1. Production monitoring to detect "risky" pieces. These rules are defined so that they compare the current production parameter values and label as "risky" those pieces that were produced with parameters outside the predefined tolerance intervals.
2. Rules that capture the scanning process and defect detection. These rules define when a defect is detected, or more precisely, the tolerance for the dimensions of the defects.
3. Rules for alerting appropriate actors in a case of defect. These are the rules that trigger appropriate mechanisms to notify the responsible personnel.

Table 3.15 Example of rule for quality control expressed in SWRL

As_Measured(?m), Heating(?h), Optimal(?o), hasOvenTemperature(?h, ?y), hasOvenTemperature(?o, ?x), greaterThan(?z, 20), subtract(?z, ?x, ?y) -> ConsideredRisky(?m, true)

For example, a rule states that if the heating temperature during production is 30°C higher than predefined optimal temperature, then the part should be considered as potentially defected. An example of uch a rule implemented in SWRL is given in Table 3.15:

3.3.3.2 Rules for Life Cycle Cost Assessment

To enable advanced control of product design and maintenance, we have created three groups of rules:

1. *Rules for enforcing customer requests*—These rules control if the current solutions according to LCC minimization correspond with the customer requirements for the product's properties.
2. *Rules for inheritance of properties from part to product*—This group of rules is used to automatically calculate the production line properties based on a set of chosen stations. For example, the surface of a product is the sum of the surfaces of its parts, while the productivity of a product is equal to the lowest part productivity.
3. *Rules for alerts in MOL stage maintenance*—These rules are used to alert service teams when the production line is not functioning as it was designed to, using automated data collection from the field.

All the rules are again implemented in SWRL. Table 3.16 gives an example of a rule that triggers an alarm to the designer in a case where his design doesn't provide a machine availability as high as the customer requested is implemented as shown in:

3.3.3.3 Rules for Knowledge-Briefs in Automated Design

Three groups of rules have been defined for the Aker use-case:

1. *Rules for maintenance of K-briefs validity and maturity*—Based on who reviewed them and when, K-briefs can be assigned a different maturity level and this set of rules automates this process.

Table 3.16 Example of rule for LCA expressed in SWRL

Customer(?c), AsDesignedProduct(?a), Alarm(?l), hasAvailability(?c, ?x), hasAvailability(?a, ?y), greaterThan(?x, ?y) -> isActive(?l, true), Text(?l, "Check Customer Required Availability"), Address(?l, "Designer")

Table 3.17 Example of rule for semantic enrichment of standards expressed in SWRL

Alarm(?l), AsDesignedMainWalk(?a), StandardMainWalk(?b), HasHorizontalLength(?a, ?c), HasHorizontalLength(?b, ?d), lessThan(?c, ?d) -> isActive(?l, true), Address(?l, "designer"), Text(?l, "check horizontal length of Main Walkways")

2. *Rules for recommending industrial standards appropriate for the project*—Relevant industrial standards are recommended via alerting system based on the location, discipline, and process.
3. *Rules for enforcing the application of semantically enriched standards*—A special ontology is developed, representing semantically enriched industrial standards, which uses an alerting mechanism when a designer decision violates certain regulations.

Table 3.17 gives an example from the third group of rules implemented in SWRL that triggers an alarm in the case that a designed main walk is not as long as required by the Norsok standard:

3.4 INTEGRATION WITH LEAP SERVICES

An important potential use of ontologies is related to the use of reasoners and how they can be used to infer new information that is not explicitly described in our structured data. Such a case in the context of LEAP is the use of the ontology together with a reasoner to identify potentially risky situations based on a number of rules that are related to the LEAP ontology. This ontology consists of an upper ontology, which acts as a common reference schema, together with three specialization ontologies for each of the LEAP use cases. The rules that are constructed encapsulate the knowledge of the domain experts for each of the three use cases and are codified using concepts and properties from the ontology.

This set of rules has been created using the Protégé tool and they are part of the LEAP ontology. These rules can be used by a reasoner to infer new concepts and provide the system with a mechanism that indicates whether or not an error is possible on a manufactured component.

To make such functionality available in the LEAP platform (in ref. to Chapter 2), it has to be made as interoperable as possible. For this reason, it was decided to create a REST interface that exposes the reasoning on the LEAP ontology over the HTTP protocol. The system implements a simple protocol by which the following happens: the interface receives data in the O-DF standard (https://www2.opengroup.org/ogsys/catalog/C14A)

format and translates them into RDF/XML statements, these statements are embedded in the LEAP ontology, a reasoner then infers new statements based on the codified in SWRL rules, and the results are returned back to the client who initiated the process.

The RDFization of the imported data happens through the instance importer mechanism. The instance importer is the mechanism responsible for taking data from the data sources layer of the LEAP platform, create instances of the LEAP ontology, and import the data in an appropriate storage mechanism. This mechanism consists of a set of rules that map arbitrary input data into the classes and properties that are defined in the LEAP ontology, together with appropriate parsers that are capable of reading and extracting information from a data source. The data serialization is assumed to be in a CSV format. Initially, some manual work is required where the columns of the CSV file are mapped to the classes and properties of the ontology using the mapping language of the importer, this information is then used every time that a new CSV arrives to extract information from it and then create instances of the ontology. It has to be noted that this manual definition of the mappings happens only once for each possible data source and then every file that arrives in the future which respects the initial structure that was mapped at the beginning will be directly instantiated and inserted in an appropriate triple store for further processing and querying. This mechanism is in the current context used to generate the RDF/XML statements from the QLM data and to feed the reasoner.

The reasoner that is used by the rule execution engine of the LEAP platform is the Pellet OWL 2 Reasoner. For applications that need to represent and reason information using OWL, Pellet is the leading choice for systems where sound-and-complete OWL DL reasoning is essential. Pellet incorporates optimizations for nominals, conjunctive query answering, and incremental reasoning. An OWL DL reasoner such as Pellet is a core component of ontology-based data management applications (http://clarkparsia.com/pellet/).

3.5 CONCLUSIONS

Knowledge structured into a semantic model creates a number of advantages, such as common terminology for all intended users, prevention of redundant entries, and easy updating. Ontology, which is becoming an increasingly popular semantic model, enables one additional advantage of high importance. Given that concepts and relations are machine understandable components, ontologies enable inference and reasoning which leads to the generation of new knowledge.

This chapter first presented the ontology structure, as well as main the methodologies and tools for ontology design. Second, we have covered in detail the process of defining and describing the LinkedDesign domain and generalization of use-cases through the definition of a collection of generic and basic concepts, their subconcepts, and relations interconnecting them, as an upper ontology. Having an upper ontology significantly reduces the time and effort invested in ontology design. Besides, to deal with the three industrial use-cases, we have created three domain specific ontologies as a specialization of the upper ontology. To formalize this specialization process, we have used a questionnaire, which enables ontology nonexperts to express their requirements for ontology functionalities. Lists of specific concepts are created based on this document and the mapping with the upper ontology is performed. The three use-cases show how different in topic and size requirements the different users can be, moving from quality control for Volkswagen-TRIMEK, LCC for COMAU, to knowledge briefs in automated design for Aker. They also show that the upper ontology can be used as is in every one of these diverse cases.

By implementing the domain knowledge and domain dynamics into one semantic model, we gain a compact overview of the domain functionalities. The same methodology can be applied for diverse specific use cases, which will result in novel and practical tools. In this chapter, we have presented ontology inference behaving as an online monitoring system. We have shown that it can behave as an automatic alert system for quality control or design errors. By recommending appropriate industrial standards, it can further act as a design decision support system. Finally, we have shown that we can design a knowledge base maintenance system by using ontology rules. At the same time, the module remains easy to interact with and convenient to integrate into various environments. The initial rules bases can be updated if necessary and the inference engine will easily adjust the inference results.

REFERENCES

[1] Gruber TR. A translation approach to portable ontology specifications. Knowl Acquisit 1993;5:199–220.
[2] Studer R, Benjamins VR, Fensel D. Knowledge engineering: principles and methods. Data Knowl Eng 1998;25:161–97.
[3] Suárez-Figueroa MC, Gómez-Pérez A, Muñoz-García O, Vigo M. gOntt, a Tool for Scheduling and Executing Ontology Development Projects; 2010.

[4] Gómez-Pérez A, Fernández-López M, Corcho O. Ontological engineering: with examples from the areas of knowledge management, e-commerce and the semantic web (advanced information and knowledge processing). New York, NY, Secaucus, NJ: Springer-Verlag; 2007.
[5] Staab S, Studer R, Schnurr H-P, Sure Y. Knowledge processes and ontologies. IEEE Intell Syst 2001;16:26–34.
[6] Pinto HS, Staab S, Tempich C. DILIGENT: towards a fine-grained methodology for DIstributed, Loosely-controlled and evolvInG Engineering of oNTologies. In: Proceedings of the 16th European Conference on Artificial Intelligence (ECAI). IOS Press; 2004. p. 393–397.

CHAPTER 4

LEAP Product and Manufacturing Design Support System

D. Cerri and S. Terzi
Polytechnic of Milan, Milan, Italy

Contents

4.1 INTRODUCTION

In the modern world, companies are increasingly required to consider the product life cycle. Companies from advanced countries have to face the low-cost pressure of emerging countries, with whom they cannot compete in terms of labor costs. Policy makers (eg, the European Commission, the UN, etc.) are also pushing for environmental life cycle considerations. In fact, in the last few years, a huge number of laws and directives have been introduced. Finally, industrial companies are changing their way of thinking. They want personalized solutions, with the lower total cost of ownership that this makes possible.

In this context, manufacturing companies have to identify the best life cycle–oriented solution to satisfy their customers' requests and survive in the global market. In fact, being able to develop ecofriendly, energy-efficient, and green products before others could give a competitive advantage in the years ahead. Designers and systems engineers are the most involved actors in the life cycle consideration: in fact, about two-thirds of total life cycle costs are fixed in the early design phases [1]. Therefore, they are the most responsible for the improvement of the product life cycle.

Taking the LEAP
http://dx.doi.org/10.1016/B978-0-12-805263-1.00004-9

Two methodologies are well known in the literature and can support the evaluation of the costs and environmental impacts generated along the whole life cycle, they are Life Cycle Costing (LCC) and Life Cycle Assessment (LCA).

LCC considers the "cradle-to-grave" costs, which can be summarized as an economic model that can be used to evaluate the alternatives for equipment and projects. Engineering details drive LCC cost numbers for the economic calculations. The economics of alternatives drives the selection process. Good engineering proposal alternatives without economic justification are often uneconomical, while good engineering with good economics provide business success. Therefore, the LCC economic model provides better assessment of the long term cost effectiveness of projects [2].

Other definitions of LCC are:

- LCC is the total cost of ownership of machinery and equipment, including its cost of acquisition, operation, maintenance, conversion, and/or decommission [3].
- LCC is the summation of the cost estimates from inception to disposal for both equipment and projects as determined by an analytical study and estimate of total costs experienced in annual time increments during the project life with consideration for the time value of money. The objective of LCC analysis is to choose the most cost effective approach from a series of alternatives to achieve the lowest long term cost of ownership. LCC is an economic model over the project life span. Usually the cost of operation, maintenance, and disposal costs exceed all other first costs many times over. The best balance among cost elements is achieved when the total LCC is minimized [4].

LCC helps to change our perspective of the mere acquisition cost with an emphasis on enhancing economic competitiveness by working for the lowest long-term cost of ownership, which is not an easy answer to obtain. Consider these typical problems and conflicts observed in most companies:

1. Project Engineering wants to minimize capital costs as the only criteria;
2. Maintenance Engineering wants to minimize repair hours and costs as the only criteria;
3. Production wants to maximize uptime hours as the only criteria;
4. Reliability Engineering wants to avoid failures as the only criteria;
5. Accounting wants to maximize project net present value as the only criteria;
6. Shareholders want to increase stockholder wealth as the only criteria.

Management is responsible for harmonizing these potential conflicts under the banner of operating for the lowest long-term cost of ownership. LCC can be used as a management decision tool for harmonizing the never-ending conflicts by focusing on facts, money, and time [2].

LCA, in contrast, is a technique to assess environmental impacts associated with all the stages of a product's life from cradle-to-grave. This concept considers the entire life cycle of a product [5]. "Cradle-to-grave" begins with the gathering of raw materials from the earth to create the product and it ends at the point when all of the materials are once again returned to the earth. LCA evaluates all of the stages of a product's life from the perspective that they are interdependent, meaning that one operation leads to the next. LCA enables the estimation of the cumulative environmental impacts resulting from all of the stages in the product life cycle, often including impacts not considered in more traditional analysis (eg, raw material extraction, material transportation, etc.). By including the impacts throughout the product life cycle, LCA provides a comprehensive view of the environmental aspects of the product or process, and a more accurate picture of the true environmental trade-offs in product and process selection. The term "life cycle" refers to the major activities in the course of the product's life span from its manufacture, use, and maintenance, to its final disposal, including the raw material acquisition that is required to manufacture the product.

Specifically, LCA is a technique that is used to assess the environmental aspects and potential impacts associated with a product, process, or service, by:

- compiling an inventory of relevant energy and material inputs and environmental releases;
- evaluating the potential environmental impacts associated with identified inputs and releases;
- interpreting the results to help decision-makers make a more informed decision [6].

LCA can help decision-makers to select the product or process that has the least impact on the environment. This information can be used with other factors, such as cost and performance data, to select a product or process. LCA data identifies the transfer of environmental impacts from one media to another and/or from one life cycle stage to another. If an LCA were not performed, then the transfer might not be recognized and properly included in the analysis because it is outside of the typical scope or focus of the product selection processes.

This ability to track and document shifts in environmental impacts can help decision-makers and managers to fully characterize the environmental trade-offs associated with product or process alternatives.

By performing an LCA, analysts can:

- develop a systematic evaluation of the environmental consequences associated with a given product;
- analyze the environmental trade-offs associated with one or more specific products/processes to help gain stakeholder (state, community, etc.) acceptance for a planned action;
- quantify environmental releases to air, water, and land in relation to each life cycle stage and/or major contributing process;
- assist in identifying significant shifts in environmental impacts between life cycle stages and environmental media;
- assess the human and ecological effects of material consumption and environmental releases to the local community, region, and world;
- compare the health and ecological impacts between two or more rival products/processes or identify the impacts of a specific product or process; and
- identify impacts to one or more specific environmental areas of concern.

However, most of the available researchers are unable to guarantee that an optimal solution will be reached; rather, in most cases, LCC and LCA are used only for simple evaluations.

The aim of this chapter is to show the development of the so-called "LEAP product and manufacturing design support system," which is able to support designers in the creation and identification of the optimal life cycle oriented solutions in terms of LCCs and environmental impacts.

In the first section of this chapter, the LCC and LCA methodologies are presented. A further section is dedicated to the so-called "Life Cycle Optimization."

The next section will describe the development of the LEAP model for the design support. Finally, the model is applied on an industrial case provided by COMAU. The last section concludes the chapter.

4.1.1 Product and Manufacturing Life Cycle Optimization

Analyzing the literature, only a few papers have been identified in the optimization of costs and/or environmental impacts along the product life cycle. This kind of optimization is called "Life Cycle Optimization," because it uses optimization methods with life cycle methodologies (ie, LCC and LCA).

An exploratory analysis of the literature has been conducted by analyzing 39 papers referring to LCC and 40 papers referring to LCA. An analysis of the results shows that only a few papers deal with the optimization issue. The percentage is about 20% for the LCC literature, which is reduced by half in the LCA literature.

The optimization methods used in the literature are linear programming, genetic algorithms, and particle swarm optimization.

In the LCC literature, the most used method is genetic algorithm (about 60%), while in the LCA literature the most used is linear programming (about 50%).

As can be seen, no contribution optimizes both cost and environmental impact in the whole product life cycle. Furthermore, life cycle optimization is only applied on some sectors. One of the sectors that is not covered is the industrial system sector, which is instead considered within the Linked-Design project.

4.2 LIFE CYCLE OPTIMIZATION IN THE LEAP

In this section, a model to optimize product LCCs and environmental impacts together, called Product Life Cycle Optimization (PLCO), is introduced. The PLCO model is first developed with genetic algorithms and compared to linear programming. Finally, it is implemented using Java frameworks.

4.2.1 Evaluation of Optimization Models and Algorithms

4.2.1.1 Linear Programming Models

Two linear programming models are used: the so-called Weighted Sum Model (WSM) and the transformation of a multiobjective problem into a single objective problem, with an objective moved under the constraints, which is called a Bi-Mono Model.

The WSM is the best-known and simplest Multi-Criteria Decision Analysis (MCDA)/multicriteria decision making method for evaluating a number of alternatives in terms of a number of decision criteria. It is very important to state here that it is applicable only when all of the data are expressed in exactly the same unit. To obviate this, we first run a single objective problem, maximizing in one run LCCs, and in the other one life cycle environmental impacts. Therefore, we are able to obtain the maximum values of life cycle costs, identified by LCC*, and life cycle environmental impacts, identified by LCA*. Then we run WSM, dividing the obtained life cycle costs and life cycle environmental impacts values with LCC* and LCA*, in order to have a sum of the two ratios.

In general, suppose that a given MCDA problem is defined on *m* alternatives and n decision criteria. Furthermore, let us assume that all of the criteria are benefit criteria; that is, the higher the values are, the better it is. Next, suppose that w_j denotes the relative weight of importance of the criterion C_j and a_{ij} is the performance value of alternative *Ai* when it is evaluated in terms of criterion C_j. Then, the total (ie, when all the criteria are considered simultaneously) importance of alternative A_i, can be denoted as $A_i^{\text{WSM-score}}$ [7], is defined as follows (Eq. 4.1):

$$A_i^{\text{WSM-score}} = \sum_j w_j \cdot a_{ij} \quad \text{for} \quad i = 1, 2, 3, \ldots, m \tag{4.1}$$

In the Bi-Mono Model, we first perform a single objective problem, which is to minimize and maximize LCCs. The minimum LCCs value is identified by *L*, the maximum is identified by *H*. Then, we run the other objective, referred to life cycle environmental impacts, while LCCs is used as a constraint, varying its value between *L* and *H*, according to a specified step. If the objective in the constraints was to maximize, then it will be a constraint of greater or equal, otherwise it will be a constraint of lower or equal.

4.2.1.2 Genetic Algorithms

Many types of multiobjective genetic algorithms exist in the literature. We decided to use Non-Dominated Sorting Genetic Algorithm 2 (NSGA-2).

NSGA-2 is one of the most popular multiobjective optimization algorithms and it has three special characteristics: fast nondominated sorting approach, fast crowded distance estimation procedure, and simple crowded comparison operator [8].

Deb et al. simulated several test problems from a previous study using NSGA-II optimization techniques and it is claimed that this technique outperformed PAES and SPEA in terms of finding a diverse set of solutions.

NSGA-2 has been demonstrated as one of the most efficient algorithms for multiobjective optimization of a number of benchmark problems [9].

The complete procedure of NSGA-2 is given below to demonstrate an implementation of elitism without using a secondary external population.

The procedure of NSGA-2 is as follows:

1. create a random parent population P_0 of size *N*. Set $t = 0$.
2. apply crossover and mutation to P_0 to create offspring population Q_0 of size *N*.

3. if the stopping criterion is satisfied, stop and return to P_t.
4. set $R_t = P_t \cup Q_t$.
5. using the fast nondominated sorting algorithm, identify the nondominated fronts $F_1, F_2, \ldots, F_k$ in R_t.
6. for $i = 1, \ldots, k$ do following steps:
   a. calculate the crowding distance of the solutions in F_i.
   b. create P_{t+1} as follows:
      – Case 1: If $|P_{t+1}| + |F_i| \leq N$, then set $P_{t+1} = P_{t+1} \cup F_i$;
      – Case 2: If $|P_{t+1}| + |F_i| > N$, then add the least crowded $N - |P_{t+1}|$ solutions from F_i to P_{t+1}.
7. use binary tournament selection based on the crowding distance to select parents from P_{t+1}. Apply crossover and mutation to P_{t+1} to create offspring population Q_{t+1} of size N.
8. set $t = t+1$, and go to Step 3. [8]

4.2.1.3 Experimental Scenarios

Three different scenarios are created to compare the three optimization methods. Scenario A has a unique optimal solution. Scenario B has more optimal solutions arranged on a Pareto front. Scenario C is equal to the second with the addition of a constraint.

This is supposed to have a generic product composed of ten subgroups. Each subgroup has two alternatives to be realized. Each alternative has data that consider all the life cycle of the product, in terms of costs and environmental impacts. Each alternative has this data input:

- *Cin*—initial cost;
- *Cmnt*—maintenance cost;
- *Cen*—energy cost;
- *Cmdpmn*—cost of manpower for maintenance;
- *BOL*—environmental impact in beginning of life;
- *MOL*—environmental impact in middle of life; and
- *EOL*—environmental impact in end of life.

The units of measurement are a generic unit of cost for LCC and a generic unit of environmental impact for LCA.

LCC is calculated as (Eq. 2):

$$\mathrm{LCC} = \sum_i \left(\mathrm{Cin}_i + \mathrm{Cmnt}_i + \mathrm{Cen}_i + \mathrm{Cmdpmn}_i\right) \cdot x_i \qquad (4.2)$$

while LCA is calculated as (Eq. 4.3):

$$\mathrm{LCA} = \sum_i \left(\mathrm{BOL}_i + \mathrm{MOL}_i + \mathrm{EOL}_i\right) \cdot x_i \qquad (4.3)$$

where x_i is a binary variable which assumes value 1 if the subgroup *i*th is used to realize the product, otherwise it assumes value 0.

The two objectives are:

Minimize the LCC;

Minimize the LCA.

In WSM, the model is written as:

Minimize

$$w \cdot (\mathrm{LCC}/\mathrm{LCC}^{\star}) + k \cdot (\mathrm{LCA}/\mathrm{LCA}^{\star}) \tag{4.4}$$

Subject to

$$\mathrm{LCC} = \sum_i (\mathrm{Cin}_i + \mathrm{Cmnt}_i + \mathrm{Cen}_i + \mathrm{Cmdpmn}_i) \cdot x_i \tag{4.5}$$

$$\mathrm{LCA} = \sum_i (\mathrm{BOL}_i + \mathrm{MOL}_i + \mathrm{EOL}_i) \cdot x_i \tag{4.6}$$

$$x_i + x_{i+1} = 1 \qquad i = 1, 3, 5, \ldots, 19 \tag{4.7}$$

$$w + k = 1 \tag{4.8}$$

$$w, k \geq 0 \tag{4.9}$$

$$x_1, x_2, \ldots, x_{20} \in (0,1) \tag{4.10}$$

The two objectives are dimensionally different: LCC has a cost dimension while LCA has an environmental impact dimension. If we want to add LCA and LCC, then we must make LCA and LCC dimensionless. Therefore, we first solve a single objective problem, maximizing one time LCA and one time LCC. We then obtain LCA★ and LCC★. We put these values in the objective function as shown earlier (Eq. 4.4). Therefore, we make LCA and LCC dimensionless and we can then sum them. Iteratively we change the values of w and k, respecting the constraint, to obtain the different solutions of the problem. The iteration starts from w = 1 and k = 0 to arrive at w = 0 and k = 1, passing through intermediate values as $w = 0.55$ and $k = 0.45$.

In the Bi-Mono Model, the model is written as:

$$\min \mathrm{LCC} = \sum_i (\mathrm{Cin}_i + \mathrm{Cmnt}_i + \mathrm{Cen}_i + \mathrm{Cmdpmn}_i) \cdot x_i \tag{4.11}$$

Subject to

$$\mathrm{LCA} = \sum_i (\mathrm{BOL}_i + \mathrm{MOL}_i + \mathrm{EOL}_i) \cdot x_i \leq \mathrm{TV} \tag{4.12}$$

$$x_i + x_{i+1} = 1 \qquad i = 1, 3, 5, \ldots, 19 \tag{4.13}$$

$$x_1, x_2, \ldots, x_{20} \in (0,1) \tag{4.14}$$

where TV is the Target Value.

In the Multi-Objective Genetic Algorithm (we have used NSGA-2) there is a chromosome (which represents the generic product) composed of ten genes (which represent the subgroups). Each gene can assume only two values (eg, gene 1 can be 1 or 2, gene 2 can be 3 or 4, ..., gene 10 can be 19 or 20). The genetic algorithm optimizes the two objectives, simultaneously creating a curve similar to a Pareto front.

Here we have used a population size of 50, a one point crossover with a rate of 0.95, and a single mutation by gene with rate 0.05.

All of those described earlier are used for each of three scenarios.

In the Test A scenario the data are arranged so that we obtain the unique solution reported in Table 4.1.

Every optimization method reaches this solution.

In the Test B scenario the models are equal to those of the previous scenario, while the data input changes. Indeed, in Test B there is no unique solution, but the solutions are distributed on a Pareto Curve. Therefore, we analyzed the behavior of the three methods (WSM, Bi-Mono Model, genetic algorithm) within Test B.

A graphical comparison is reported in Fig. 4.1.

As you can see, NSGA-2 provides a set of nondominated solutions that are more complete than the other two optimization methods. WSM finds fewer nondominated solutions than NSGA-2. This is probably caused by the chosen step (we have used a step of 0.05). This is the major disadvantage of WSM. The advantage is that WSM found solutions that are surely on a Pareto Curve. This allows a comparison to be made between NSGA-2 solutions and WSM solutions. It is possible to observe from the graph that NSGA-2 solutions, corresponding to WSM solutions, are surely on a Pareto Curve. In contrast, Bi-Mono finds a good number of solutions, but many of them are dominated.

Table 4.1 Solution of Scenario A

Product Subgroups	1	3	5	7	9	11	13	15	17	19
Min LCC	589									
Min LCA	430									

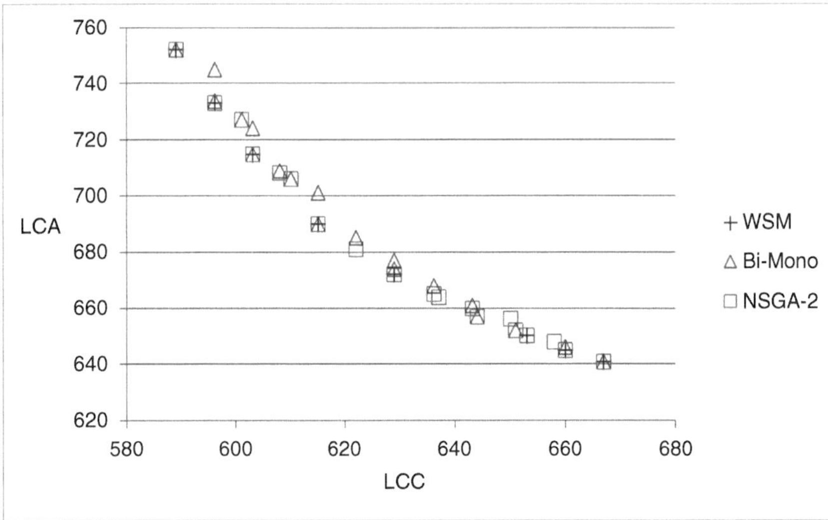

Figure 4.1 ***Graphical comparison of the solutions (Scenario B).***

In the Test C scenario the data input and models are equal to the second scenario, with the addition of a column of values which must respect the following constraint, which is added in the models (Eq. 4.15):

$$\sum_i g_i \cdot x_i \leq G \tag{4.15}$$

where g_i is a generic value of the ith subgroup, while G is the threshold value.

A graphical comparison is reported in Fig. 4.2.

The graphical comparison is very similar to that in Scenario B. NSGA-2 provides more nondominated solutions than WSM. Bi-Mono, instead, provides some dominated solutions. Therefore, we can affirm that NSGA-2 is a robust and reliable optimization method.

Summarizing, results of this analysis are:

- NSGA-2 provides a larger number of nondominated solutions than the other two optimization methods;
- NSGA-2 is a robust and reliable optimization method (it provides optimal solutions: we can say this by comparing NSGA-2 solutions with WSM solutions);
- WSM finds optimal solutions, but fewer than NSGA-2;
- Bi-Mono finds a good number of solutions, but some of these are dominated.

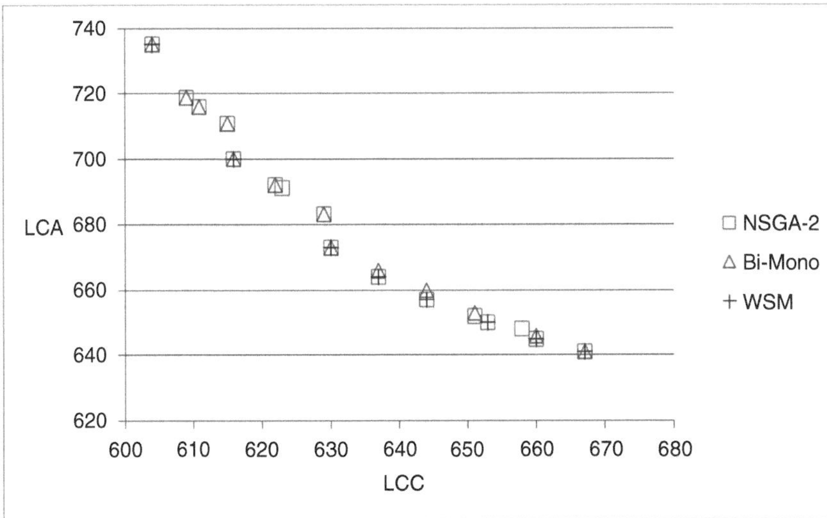

Figure 4.2 ***Graphical comparison of the solutions (Scenario C).***

4.2.2 Life Cycle Optimization Model Implementation in the LEAP

In this paragraph, the model previously presented is implemented in a Java application. In detail, the so-called "Front-End" (user interface) is developed by ZK [10], that is an open source Java web framework. The so-called "Back-End" is created into an object-oriented Java based framework, JMetal [11]. JMetal is a library to develop, experiment, and study metaheuristic for solving multi-objective problems. Fig. 4.3 shows the model framework.

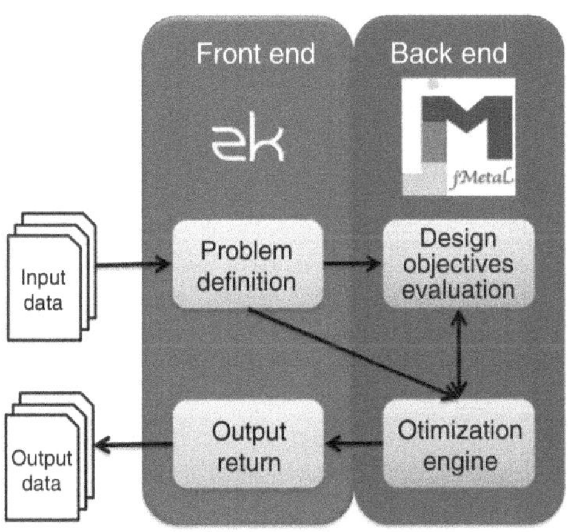

Figure 4.3 ***Model implementation framework.***

4.2.2.1 Front End

The front-end side of the model includes the Problem Definition module and the Output Return module. The problem to be defined requires various information, which comes out from the customer requests and from the designer needs. Furthermore, the output of the algorithm should be clearly and understandable visualized. Thus, it is indispensable to build a User Interface (UI) that allows and helps the designer to set in a simple way all the parameters required and permits to display the achieved output.

ZK, an open source Java framework, is used due to its easiness and rapidity to create a UI, thanks also to ZK components. They can be configured to fit the developer desires.

The first objective of the developing the UI is to design an index window with specified buttons and forms, in order to set the parameters of the problem. The PLCO index UI is divided into two main areas: Objective Functions Definition and Design Space Definition (see Fig. 4.4).

In the Objective Functions Definition area, coherently with the model framework, it is possible to:

- Choose the objectives to evaluate in the optimization model ticking the checkbox;

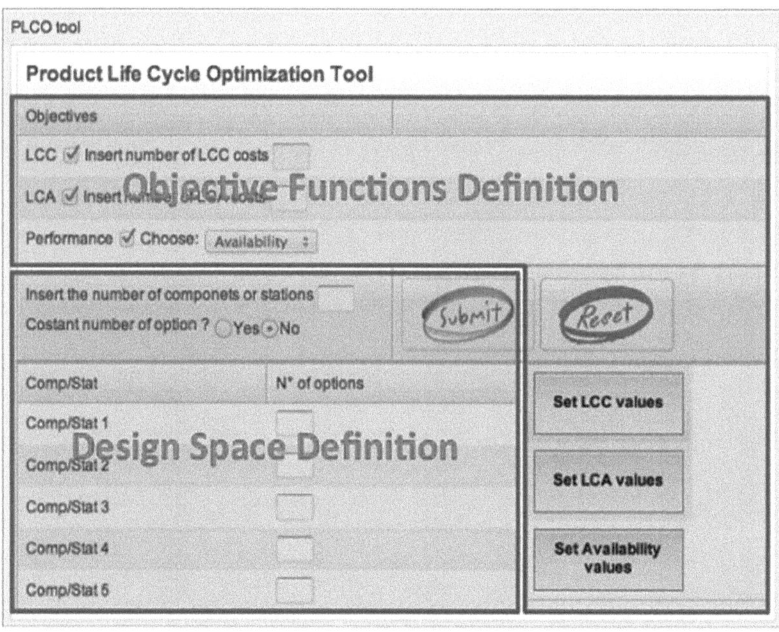

Figure 4.4 *Division of the index windows based on the model framework.*

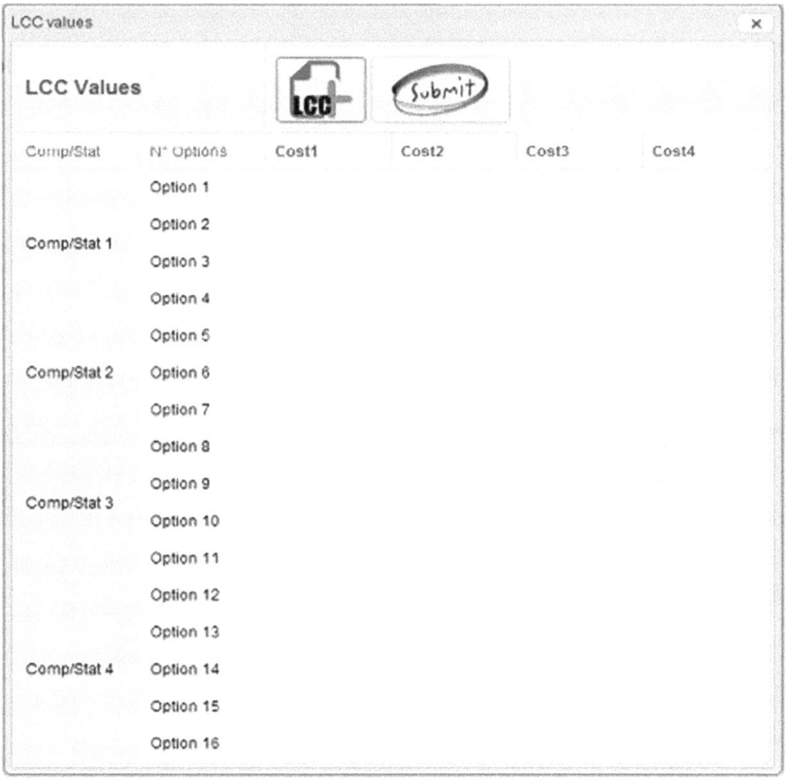

Figure 4.5 ***Table for the LCC data matrix insert.***

- Define how many components compose the objective functions filling the textbox;
- Insert the data matrix containing the LCC, LCA, and Performances values by clicking the dedicating button and selecting the file (Fig. 4.5 displays the data insert windows).

In the other area, the Design Space Definition, it is possible to:

- Define the number of variables filling the textbox (components or stations);
- Define the number of options for each variable filling the textbox.

All of the parameters and the data matrices that are inserted are checked and validated in order to avoid errors. If all of the data are correct, then the interface shows a hidden button that allows the algorithm execution (Fig. 4.6).

The button invokes a method in the back-end side.

The second objective of the front-end side is to implement the Output Return module. Thus, after the algorithm has been executed, it is necessary

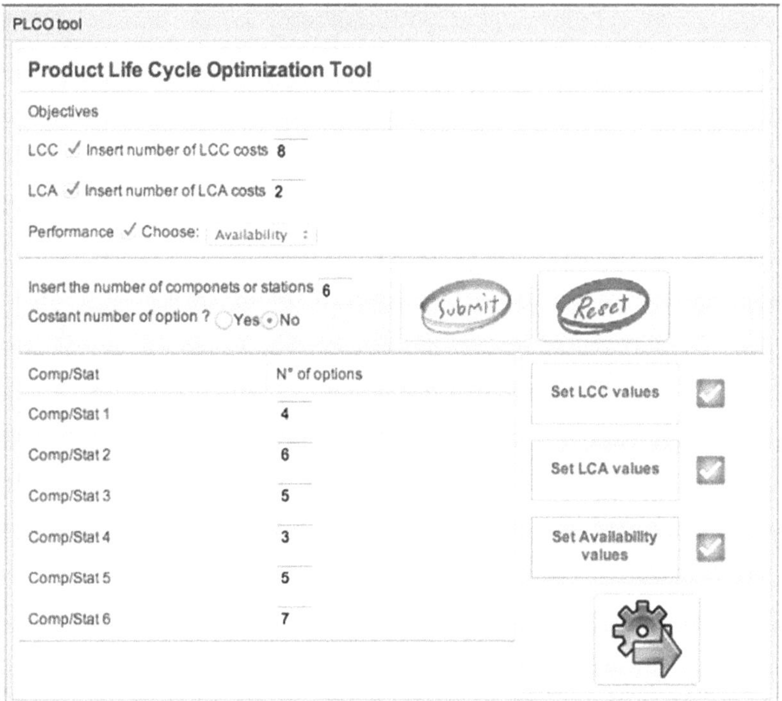

Figure 4.6 *The execute button showed in the index windows.*

display the achieved results in the browser. Fig. 4.7 shows the Output Return screen, which reports:

- Values of the main parameters;
- Values of the objective functions; and
- Alternative solutions (a Multi-Criteria Decision Making Approach is under development).

Moreover, it is possible to store the results in an Excel file by clicking the save button and the Pareto front graphs can be displayed by clicking the graph button.

The last step is to make the UI responded to the users. The approach introduced is to control the UI component directly. This approach can be classified to Model-View-Controller (MVC) design pattern. This pattern divides an application into three parts:

- The View means UI. The page (a XML-formatted language), which contains ZK components, represents this part. A user's interaction with the components triggers events to be sent to the controllers.

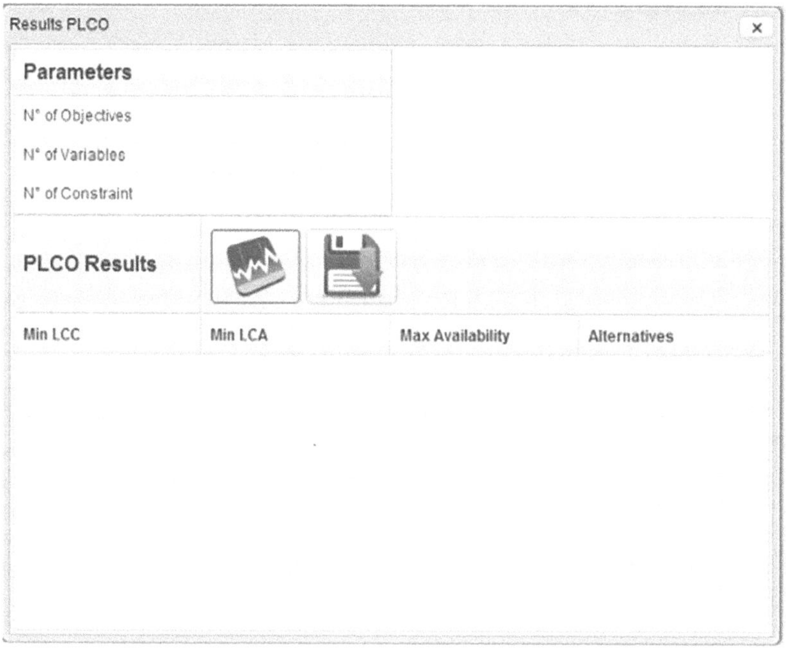

Figure 4.7 ***PLCO output visualization.***

- The Controller plays the role of coordinator between View and Model. It receives events from View to update Model and retrieve data from Model to change View's presentation.
- The Model consists of application data and business rules. In the proposed model it corresponds to the logic inside the back-end side.

When a user interacts with a component (eg, click a button) on the page, the user's action triggers an event. This event is sent to the controller and invokes a corresponding event listener method. The event listener method usually executes business logic or accesses data, then, it manipulates ZK components. A component's state change in an event listener is reflected in its corresponding UI.

This ZK pattern allows us to control the UI via a controller at the server-side; the controller is, therefore, the main extension point to integrate any Java library or framework. To integrate ZK with other frameworks (JMetal), it is necessary to write a controller code in order to use classes of back-end systems.

4.2.2.2 Back End

The back-end side of the model includes the Design Object Evaluation module and the Optimization Engine Module of the PLCO problem, by using the NSGA-2 algorithm in JMetal. JMetal [11] provides a rich set of classes that can be used as the building blocks of multiobjective techniques. Furthermore, it contains a number of state-of-the-art algorithms (included NSGA-2) and a set of quality indicators that not only allows newcomers application to study the basic principles of multiobjective optimization with metaheuristics but also their application to solve real-world problems like PLCO. JMetal is chosen to develop the algorithm because it is simple and easy to use, portable, flexible, and extensible.

The Unified Modeling Language (UML) describes the architecture and components of JMetal, used for the PLCO. A UML class diagram, showing the main components and their relationships is depicted in Fig. 4.8. The PLCO problem is built on the basic architecture of JMetal, which relies on an Algorithm to solve a Problem, using one SolutionSet and a set of Operator objects.

As shown in the previous class diagram (Fig. 4.8), the basic components of JMetal framework are:

- solution encodings;
- operators;
- problems; and
- algorithms.

It is essential to ensure the correct implementation of the PLCO problem by establishing the proper set of these components.

4.2.2.2.1 Encoding of Solution

One of the first decisions that has to be taken when metaheuristics methods are used is to define how to encode or represent the tentative solution of the problem to solve. Representation strongly depends on the problem and determines the operation that can be applied. Thus, selecting a specific representation has a great impact on the behavior of metaheuristics and, consequently, on the obtained results. A Solution is composed of a set of Variable objects, which can be of different types, plus an array to store the fitness value. Effectively, the variable and solution in the PLCO problem stand for sets of suitable subgroups of machines; hence, it must be an integer.

4.2.2.2.2 Operators

Metaheuristic techniques are based on the modification or on the generation of new solutions from existing ones by means of the applications

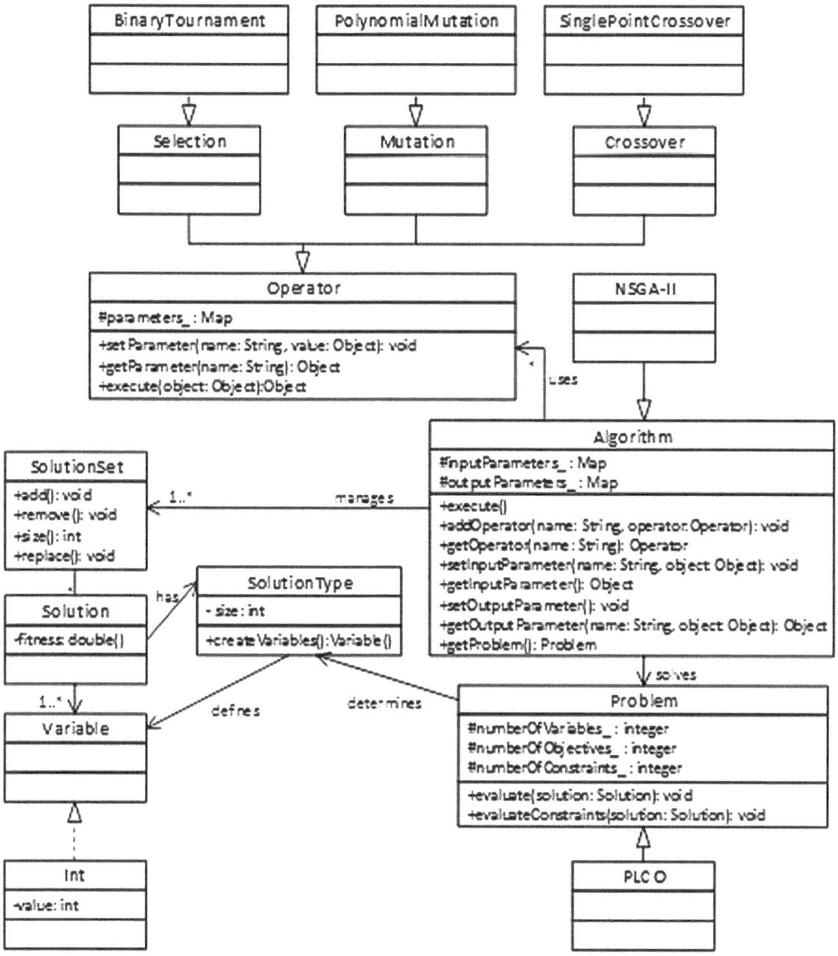

Figure 4.8 ***JMetal class diagram of PLCO.***

of different operators. The NSGA-2 makes use of crossover, mutation, and selection operators to modify solutions. The framework includes a number of different operators, thus it is relevant to select the adequate techniques in order to reduce the time complexity of the algorithm. The chosen types of operators are:

- *Crossover*—Single-point crossover;
- *Mutation*—Polynomial Mutation; and
- *Selection*—Turnover Selection.

4.2.2.2.3 Problems

In JMetal, all of the problems inherit from the class Problem. This class contains two basic methods: evaluate() and evaluateConstraint(). Both methods receive a Solution, representing a candidate solution to the problem: the first evaluates it while the second determines the overall constraint violation of the solution. All of the problems have to define evaluate() method, while only problems having constraints have to define evaluateConstraint(). In JMetal, the problem defines the allowed solutions types that are suitable to solve it.

4.2.2.2.4 Algorithm

The last core class is Algorithm, a conceptual class included in the framework, which must be inherited by the metaheuristic. The algorithm chosen to solve the PLCO problem is the NSGA-2. Given that every metaheuristic developed in JMetal, the NSGA-2 extends the class Algorithm and inherits from it the conceptual method execute() (Fig. 4.8), this is called to run the algorithm. This class uses the logic of the Engine Optimization module.

To ensure the correct execution of the algorithm, a rigorous setting of the parameters of the framework is mandatory. Given the structure of JMetal, the classes to set up the problem are PLCO_3obj and NSGAII_PLCO_run.java. Within the problem, all of its features have to be defined:

- *Number of variables*—number of decision variables which represent different operations, stations, or machines that compose the system;
- *Number of objectives*—number of functions to optimize;
- *Number of constraint*—number of conditions to respect;
- *Variables upper and lower limit*—bounds of the variables; and
- Data matrix containing Costs, Environmental Impacts and Performance values.

This information is passed to the problem through a Java object (object4Plco.java) filled in the controller of the UI.

All of the typical parameter values that are necessary for running the NSGA-2 can be set up in the class NSGA_PLCO_run.java. One of the main difficulties that a user faces when he applies an evolutionary algorithm is to decide an appropriate set of these parameter values:

- *Parameters related to the generation of the problem instance*—the problem constructor needs to be created from two values; as mentioned earlier, it is necessary to configure the class problem, inserting a sting containing the solution type and a Java object containing all the indispensable parameters;

- *Population size (N)*—population of candidate solutions, which depends on the definition of the problem;
- *Number of generation (T)*—number of maximum iterations;
- *Parameters related to crossover*—the crossover operator and the crossover probability must be set up;
- *Parameters related to mutations*—the mutation operator and the mutation probability must be set up; and
- Parameters related to selection the selection operator must be set up.

4.3 LIFE CYCLE OPTIMIZATION IN THE COMAU CASE

The previous model was tested in order to check its own soundness and is then applied on a real industrial case. COMAU, a leading manufacturing company active in the automotive sector, provided the case. The company designs in detail and realizes production and assembly equipment for car components (eg, engine assembly line, body production shops, etc.).

The model has been applied in detail to the real case of a fraction of an assembly line for a small car diesel engine. The layout is reported in Fig. 4.9. The fraction of the line is comprised of five stations, which realize the following operations:

- *OP180*—silicon coating is applied;
- *OP190*—base is assembled;

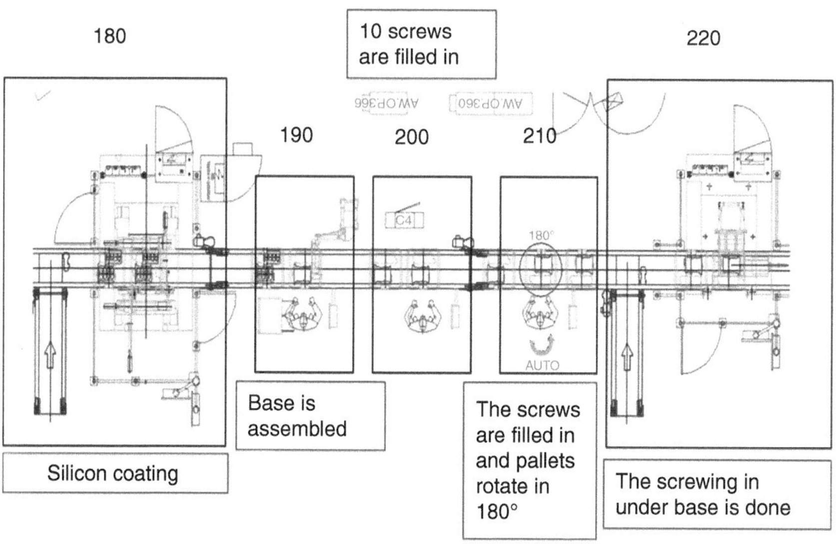

Figure 4.9 ***Reference layout of the line.***

- *OP200*—10 screws are filled in;
- *OP210*—10 screws are filled in and pallets are rotated of 180°; and
- *OP220*—screwing in under the base is done.

All of these locations can have several alternatives, including automatic (a), semiautomatic (saut), or manual (m) stations.

The costs and environmental impact related to the stations, and used for LCC and LCA analysis, are:

- Cin = initial cost, the acquisition cost of the station;
- Ce = electric energy cost;
- Cric = spare parts cost;
- Cop = labor cost (the number of workers depends from the type of station: 1 worker for 1 manual or semiautomatic station, 0.2 worker for 1 automatic station);
- Ccon = consumables cost (eg, oil and grease);
- Cair = air cost;
- Cmo = preventive maintenance cost;
- Cmorip = corrective maintenance cost;
- EIst = environmental impact of the station;
- EIel = environmental impact of electric energy; and
- A = availability of the station.

The analysis of life cycle environmental impacts is very limited because this is the very first application of this study within the company. Until now, the only data that have been retrievable are: (1) the environmental impact of the station, in terms of environmental impacts related to materials used in the building of the station, and (2) the environmental impact of the electric energy consumed by the station. Clearly, these data are not sufficient to conduct and perform an LCA analysis, even in a simplified form. However, this is an effort to introduce to the company the concept of LCA and environmental impact.

The time horizon is 10 years, while the Discount Rate (or Bank Rate) is 1.5%. The model has two objective functions, one that minimizes the product LCCs (Eq. 4.16) and one that minimizes the environmental impact during the whole life cycle (Eq. 4.17). The model has two types of constraints: the availability of the fraction of the assembly line must be greater than 0.95 (Eq. 4.18), and all of the locations must have only one station, whether it is automatic, semiautomatic, or manual (Eqs. 4.19, 4.20, 4.21, 4.22, and 4.23). Below, the model is written in analytical form:

$$\min \sum_{i=1}^{30} \left(\text{Cin}_i + \text{Ce}_i + \text{Cric}_i + \text{Cop}_i + \text{Ccon}_i + \text{Cair}_i + \text{Cmo}_i + \text{Cmorip}_i\right) \cdot x_i \tag{4.16}$$

$$\min \sum_{i=1}^{30} \left(\mathrm{EIst}_i + \mathrm{EIel}_i\right) \cdot x_i \tag{4.17}$$

Subject to

$$\sum_{i=1}^{6} \mathrm{A}_i x_i \cdot \sum_{i=7}^{12} \mathrm{A}_i x_i \cdot \sum_{i=13}^{18} \mathrm{A}_i x_i \cdot \sum_{i=19}^{24} \mathrm{A}_i x_i \cdot \sum_{i=25}^{30} \mathrm{A}_i x_i \geq 0.95 \tag{4.18}$$

$$\sum_{i=1}^{6} x_i = 1 \tag{4.19}$$

$$\sum_{i=7}^{12} x_i = 1 \tag{4.20}$$

$$\sum_{i=13}^{18} x_i = 1 \tag{4.21}$$

$$\sum_{i=19}^{24} x_i = 1 \tag{4.22}$$

$$\begin{aligned} &\sum_{i=25}^{30} x_i = 1 \\ &x_i \in \{0,1\}\, i = 1,2,\ldots,30 \end{aligned} \tag{4.23}$$

where the various costs, environmental impact and availabilities are as described earlier and x_i is a binary variable.

The algorithm returns a series of information, related to each point brought up in Fig. 4.10 (A, B, C, and D), and reported in Table 4.2. The information includes the nondominated solution, related to costs and

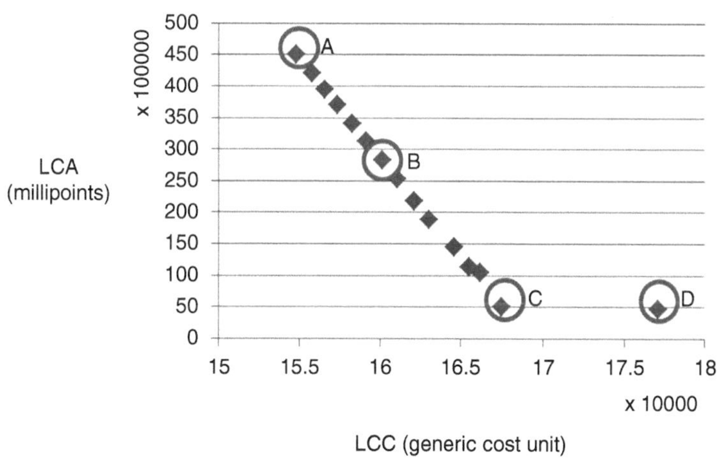

Figure 4.10 ***Graphical results of the COMAU case.***

Table 4.2 Results of the COMAU case

	Min LCC (unit cost)	Min LCA (mill points)	Stations
A	154823	45048190	2(aut) 7(aut) 13(aut) 20(aut) 26(aut)
B	160104.7	28230730	2(aut) 9(aut) 18(m) 20(aut) 30(m)
C	167493.8	5062750	6(m) 12(m) 18(m) 24(m) 30(m)
D	177139.6	4627975	6(m) 12(m) 18(m) 23(saut) 30(m)

environmental impacts generated along the whole life cycle, and the sequence of stations that permits the achievement of the previous values.

Data coming from the analysis performed with the PLCO model are therefore evaluated by some experts of COMAU in order to understand if the values returned by the tool can have sense. Experts compared these results with some of the results of their previous analyses and concluded that the algorithm returned optimal solutions

4.4 CONCLUSIONS

This chapter aimed to show the development and implementation of the so-called "LEAP model for design support." Starting from the analysis of the companies' operating context and from the analysis of the literature, it has been demonstrated how product life cycle could be a key leverage in order to compete in the global market. Some methodologies about the evaluation of costs and environmental impacts along the whole life cycle of a product, which are respectively LCC and LCA, were already well known in the literature by the 1960s. However, a gap is identified in the combination of life cycle methodologies with the optimization methods, which could enable the creation and identification of optimal life cycle oriented solutions. Within the LinkedDesign project, a model for the so-called "Life Cycle Optimization" was developed and implemented. The LEAP model for design support is based on genetic algorithm, which was the optimization method that better suited the problem faced in the project. The LEAP model for design support was developed in two components, using Java Web Frameworks: the front-end, using ZK, and the back-end, using JMetal.

Finally, the model was applied to a case provided by COMAU.

It has been evaluated via questionnaire by COMAU personnel. It satisfies some of the strategic requirements of COMAU, such as the capability to reduce costs and environmental impacts along the whole life cycle, it is also able to analyze different alternatives. Some criticalities are instead identified about the simplicity to use the model.

More information about the questionnaire is reported in Chapter 8, under the paragraph about the COMAU case.

REFERENCES

[1] Blanchard BS. 1991 Tutorial notes annual reliability and maintainability symposium. Evans Associates 804 Vickers Avenue, Durham, NC 27701; 1991.
[2] Barringer HP. A life cycle cost summary, ICOMS; 2003.
[3] Society of Automotive Engineers. Reliability and maintainability guideline for manufacturing machinery and equipment. Warrendale, PA: Society of Automotive Engineers; 1999.
[4] Landers RR. Product assurance dictionary. Marlton, NJ: Marlton Publishers; 1996.
[5] Curran MA. Environmental life cycle assessment. New York: McGraw-Hill; 1996.
[6] Scientific Applications International Corporation. Life cycle assessment: principles and practice; 2006, p. 80 (Technical Report).
[7] Triantaphyllou E. Multi-criteria decision making: a comparative study. Dordrecht, The Netherlands: Kluwer Academic Publishers; 2000.
[8] Deb K, Pratap A, Agarwal S, Meyarivan T. A fast and elitist multi-objective genetic algorithm: NSGA-II. IEEE Trans Evol Comput 2002;6(2):182–97.
[9] Murugan P, Kannan S, Baskar S. NSGA-II algorithm for multi-objective generation expansion planning problem. Electr Pow Syst Res 2009;79:622–8.
[10] http://www.zkoss.org.
[11] Nebro AJ, Durillo JJ. jMetal 4.3 user manual. Computer Science Department of the University of Malaga; 2013.

FURTHER READING

[1] Okasha NM, Frangopol DM. Lifetime-oriented multi-objective optimization of structural maintenance considering system reliability, redundancy and life cycle cost using GA. Struct Saf 2009;31:460–74.

CHAPTER 5

LEAP Collaboration System

K. Kristensen, J. Krogstie, D. Ahlers and M. Mehrpoor
NTNU, Trondheim, Norway

Contents

5.1 INTRODUCTION

In essence, most of the major achievements of humankind are the result of collaborative efforts. Even the great scientists of the past—admired for their breakthrough accomplishments—often did not develop their ideas in isolation, but rather through discussing and debating those ideas with their peers at scientific seminars and conferences. Although many of the reasons for collaborating with others remain the same, the number of ways that engineers and other knowledge professionals today collaborate and interact with each other and with customers and partners is ever increasing. The list includes, but is not limited to, different forms of formal meetings, such as design reviews, informal interactions by the water cooler, emails, video-conferences and telepresence sessions, chat sessions, collaboration using shared workspaces, small and large workshops, conference calls, and social collaboration software. These different forms of collaboration are all commonplace in many organizations. Nearly everything we do as knowledge professionals involves collaboration or interaction of some kind. Moreover,

Taking the LEAP
http://dx.doi.org/10.1016/B978-0-12-805263-1.00005-0

an increasing portion of tools and platforms are now also available through smartphones and tablets—often through dedicated apps, providing powerful collaboration functionalities to mobile professionals wherever they are.

Executives, engineering managers, specialists, and subject matter experts now often spend up to 60–80% of their time on different types of interactions; see, for example, Williams et al. [1]; Butler et al., [2]; Beardsley et. al. [3]—or Morello and Burton [4]. Most companies tend to address collaboration in a fragmented, non-disciplined, and ad hoc manner, despite the fact that the majority of the time of many engineering managers and knowledge professionals is spent on collaborative activities. This is a paradox—despite the fact that collaboration in all its forms is so commonplace, collaborative activities are often not managed, led and supported the way companies manage, lead and support individual activities. In many ways, this represents a lost opportunity.

New methods, tools, and frameworks now exist to extend the power of analytics into collaboration. The use of such tools and frameworks in combination with the close management of a set of distinct success factors for collaboration can enhance the performance of multidisciplinary, split-location engineering teams by providing insights, evaluation criteria, and clear action points to drive continuous improvement in these teams.

Today's engineering operations are very knowledge-intensive, and the ability to translate knowledge and expertise into profitable products and services is essential for international competitiveness. A study on collaborative advantage by Hansen and Nohria [5] concludes that:

> *Collaboration is perhaps the key skill required of today's organizations. And it will only become more pivotal in the future as boundaries among markets, industries, companies and countries continue to blur. Developing strategies and techniques for building collaborative advantage is considered to become the primary challenge for all companies and the primary differentiator of great ones.*

Effective and efficient collaboration is required on multiple levels in any organization, and the ability to collaborate closely and successfully by sharing knowledge and jointly developing new products and services is emerging as a key source of competitiveness. Corporations that are able to establish, stimulate, and support collaboration in flexible, dynamic patterns that are well aligned with business opportunities can leverage their combined set of dispersed resources available in their network of subsidiaries, global divisions, suppliers, customers, and partners around the globe. Furthermore, the portion of collaborative activities is steadily increasing as (1) products and services are growing in complexity; and (2) product/service innovation processes, value networks, and systems are becoming increasingly global. For example, new technologies used in the manufacturing processes produce information of potential interest not only for the production department (the team that will normally analyze such information), but also for other departments across the company. As a result, new potentially valuable links among departments emerge.

Technological advances constitute key enablers for better collaboration, but technology is useless without a suitable (design) process and the right skill-set among collaborators. This chapter documents the research results from LinkedDesign on how to work systematically with improving collaborative performance in engineering teams. In particular, this chapter covers the LEAP collaboration system, a set of models, approaches, diagnostic tools, and best practices enabling effective and efficient collaboration within multidisciplinary split-location teams working on large-scale engineering projects.

5.1.1 Objectives—Collaboration in the Context of Engineering

The goal of the research task "Concurrent engineering requirements and common design methodology basis" was to collect and analyze integrated product development and concurrent engineering requirements (design and production data), and to analyze lean product development principles based on semantics and front end value drivers. The analysis includes requirements for project portfolio management based on semantic activity space requirements, context-specific dashboard requirements for continuous monitoring and dynamic optimization and decision support and prioritization requirements.

The goal of the research task "Lean engineering collaboration system diagnostics and optimization" was to establish a solution for documentation and communication of best practice, securing that recommended solutions are available and used. This includes evaluation of engineering tools, such as Computer-Aided Design (CAD), Computer-Aided Engineering (CAE), Enterprise Research Planning (ERP) and so on, as knowledge sources, and an analysis of innovation tools and external innovation interfaces. New concepts and principles for lean engineering collaboration have been explored, as well as new diagnostics frameworks for lean engineering collaboration system analysis and optimization.

The aforementioned tasks formed an integral part of the work package on "Collaborative Environments and UI Concepts for Context-driven Engineering," where the Virtual Obeya (Chapter 7) was one of the main outputs. Common for all the research tasks in this work package is a focus on collaboration as an enabler of effective and efficient teamwork—particularly in split-location engineering teams. A combination of diagnostics tools and frameworks, surveys, workshops, and end user consultations has developed new insights on how to improve collaborative performance through identifying and addressing different sources of waste and collaboration barriers. Specifically, the chosen lean engineering approach identifies ways of reducing waste in collaboration processes. As part of this work, an evaluation of knowledge sources is given, together with ways that support knowledge creation in a Virtual Obeya (see also Chapter 7).

End users will use LinkedDesign to explore smart ways of integrating task identification, task execution, and user collaboration workflows, improving overall project effectiveness and efficiency. This is achieved through pinpointing interfaces between Knowledge Based Engineering (KBE) and related user collaboration activities, focusing on improving in-context workflows.

This work is strongly related to the conceptualization of the Virtual Obeya as a collaborative work environment where it is possible to explore and pilot in-context collaborative work in environments developed for effective and efficient knowledge sharing, visualization, and joint decision making. The research indicates that businesses can achieve significant productivity gains from improving the way that they collaborate because collaboration in all its forms is such a dominant category of activities for knowledge professionals. However, because it comes in so many forms and because collaboration involves communication, culture, status, organization, workplaces, language, processes, technologies, habits, incentives, and a range of other aspects, improving it necessarily requires a broad, multidisciplinary approach. Approaches focusing on pure technology development and deployment are considered far less effective than broader, systemic approaches to improving collaborative performance.

Specific objectives include improving project execution performance (effectiveness and efficiency) and creating new opportunities through the development, introduction, promotion, and "anchoring" of new in-context collaboration principles and methods. Finally, objectives include introducing high-performance collaboration practices that contribute to sustainable competitiveness, improved quality, project execution excellence, innovation capabilities, and productivity improvement. In this context, reuse of knowledge and information for new purposes across multidisciplinary is of key importance.

As shown by Williams et al. [1], engineers spend a lot of time interacting with systems or abstract data, and team interactions and different forms of collaboration do not happen in an information vacuum—rather the opposite. Engineering teams collaborating toward common goals spend a lot of time interacting with different tools and data sources; some are consulted individually and later brought into the collaborative activities, while others are consulted collaboratively, during a collaborative session. The following section contains an overview of lean engineering collaboration and it will describe how identified sources of waste can be reduced in collaboration.

5.2 LEAN ENGINEERING COLLABORATION

Collaboration often involves the joint use of data from different data sources. In engineering companies, a large number of different tools is used by different people. The data sources that we have selected to investigate for integration issues are of the types found particularly relevant in the use cases of the LinkedDesign project. Relevant data sources for each industrial use case (in offshore engineering, automotive, and robot manufacturing) have been identified, in particular linked to user stories from the different cases. For a more concrete example, the case in offshore engineering (see Chapter 8) relates to platform development using KBE, as described by La Rocca [6]. KBE requires close collaboration between engineers in different domains, handling data in various data formats, and a well-structured knowledge acquisition technique followed by transparency and traceability in design automation. As one example, the KBeDesign™ software is used by engineers with different roles in a product's lifecycle. The goal of KBeDesign is to visualize product knowledge and lifecycle information in an easily accessible way, enhance data integration, and improve collaboration between the engineers involved.

In this project, based on the core needs of the use cases, we have focused on the following concrete tools and tool types in the assessment.

- *Office automation*—Much data relevant for engineers and other business professionals is developed and resides in office automation tools such as Excel, see Hermans [7].
- *Computer-Aided Design (CAD)*—This is primarily used by the designers and some of the engineers, often in an early stage of product development.

- *Knowledge based Engineering (KBE)*—This has its roots in applying AI techniques (especially LISP-based) on engineering problems. In La Rocca [6] four approaches/programming languages are described: IDL, GDL, AML, and Intent!, all being extensions of LISP. In LinkedDesign, one particular KBE tool is used—KBeDesign. The KBeDesign is an engineering automation tool that was developed for oil and gas offshore platform engineering design and construction, and is built on top of a commercial KBE application (Technosoft's AML). In the use case, there are two important data sources—the representation of the engineering artifacts themselves, and the way that the engineering rules are represented (in AML) as part of the code.
- *Product Lifecycle Management (PLM/PDM)*—This is the process of managing the entire lifecycle of a product from its conception, through design and manufacture, to service and disposal. Whereas CAD systems focus primarily on the early phases of design, PLM attempts to take a full lifecycle view. PLM intends to integrate people, data, processes, and business systems, and it provides a product information backbone for companies and their extended enterprise. There is typically a core group of people creating information for such tools, and a vast group of people using this information.
- *Enterprise Research Planning (ERP)*—This is defined as a "Framework for organizing, defining, and standardizing the business processes necessary to effectively plan and control an organization so the organization can use its internal knowledge to seek external advantage" [8]. ERP systems have traditionally focused on internal process integration of well-understood business functions, such as sales, production, and inventory management [9].

An important aspect of the case is to reduce waste in lean engineering processes, based on an overview of ten forms of wasted effort in collaboration by Manyika et al. [10]. In LinkedDesign, the use case partners and other project partners have prioritized the waste areas, and we have used this input to come up with the following list of waste to be avoided:

- *Searching*—time spent searching for information.
- *Undercommunication*—excessive or not enough time spent in communication.
- *Misunderstanding*—between different people, typically having different roles.
- *Interpreting*—time spent on interpreting communication or artifacts.
- *Waiting*—delays due to reviews, approvals and so on.
- *Extra processing*—excessive creation of artifacts or information.

An analysis of the reuse of data from the varying data sources and combining these has been done and we summarize the main points here relative to factors for waste reduction in lean engineering:

Excel data

- *Searching*—When Excel is used, there is often data in a number of different Excel-sheets developed by different people, and it can be hard to know that one has the right version.
- *Undercommunication*—There is no explicit data model, thus the interpretation of data might be based on labels only, which can be interpreted differently by different persons. A number of (calculation) rules are captured in Excel-sheets without being visualized.
- *Misunderstanding*—Due to potentially different interpretation of terms, misunderstandings are likely.
- *Interpreting*—Since the meaning of data is undercommunicated, the time for interpretation might be long.
- *Waiting*—If data must be manually transformed to another format to be usable this might be an issue.
- *Extra processing*—Due to the versatility of tools such as Excel, it is very easy to represent additional data and rules, even if they are not deemed useful by the organization.

CAD data

- *Searching*—Finding the relevant CAD data can be made easier by linking it to enterprise tools such as PLM tools.
- *Undercommunication*—In CAD tools, there might be limitations in the representation of underlying design rules and process information.
- *Misunderstanding*—Due to the number of assumptions that are included in a product design that is not represented explicitly, there might be misunderstandings at a later stage.
- *Interpreting*—A number of tools exist to do different types of analysis, which might make it easier to support interpretation of the product model.
- *Waiting*—If other than designers and engineers need information from CAD tools, or need to have changes done at a later stage, they might be dependent on the availability of the engineer to do the changes.
- *Extra processing*—Since CAD tools store the geometry on the instance level, reuse for variants of products, for example, might take extra time.

KBE data

- *Searching*—Representing all of the rules in the KBE-system is useful but they are to a limited degree structured, for example, relative to how the rules influence each other.

- *Undercommunication*—Since AML-rules are accessible as code only, it can be hard to understand how different design decisions are enforced.
- *Misunderstanding*—This can result from not having access to the rules directly.
- *Interpreting*—Additional time might be needed for interpretation for the aforementioned reason.
- *Waiting*—Getting support quickly for updating rules (if necessary) can be an issue. The use of dynamic instantiation can alleviate this. On the other hand, one needs people with specific skills to add or change rules.
- *Extra processing*—We might need to represent rules differently to be useful in new situations. On the other hand, this can be addressed if the abstraction mechanisms are used well.

PDM/PLM data

- *Searching*—Large models and a lot of extra data might make it difficult to get an overview and find all of the relevant information. On the other hand, since one has a common data model, it should be easier to find all data relevant for a given product.
- *Undercommunication*—Since extra data has to be added up front for the use later in the product life cycle, it is a danger that not all necessary data is added (or is added with poor quality), which can lead to the next two issues:
  - *Misunderstanding*—Can be a result of undercommunication.
  - *Interpreting*—When engineers and other groups need to communicate, one should also be aware of possible misunderstandings, given that it seems to be hard to learn these tools if you are not an engineer. Also given that only a few people are actually adding data a lot of people need to interpret this data without actively producing it.
- *Waiting*—It can be a challenge when a change is done for this to be propagated to (eg, ERP systems and supplier systems). For some type of data this propagation can be made automatic.
- *Extra processing*—Necessary to add data up front. Can be a challenge when you need to perform changes, to have the data produced in earlier phases updated.

ERP data

- *Searching*—Enterprise systems tend to cover more and more business functions, connecting back-office operations with front-office systems, as well as moving toward real-time tracking and monitoring of operations, see Lyons [11]. The result is that the amount of information within ERP systems is increasing. This creates improved possibilities for supporting decision making, but also poses a challenge as to how to present data in a way that is useful for those making the decisions. On the other hand, since one has a common data model, it is potentially easier to find relevant data. ERP systems are known for poor search interfaces. Generally, ERP systems are good for precise queries and they are less good for vague exploratory queries.

- *Undercommunication*—A possible effect of the work-benefit disparity issue [12] is that in PLM systems some data might not be available in good quality.
- *Misunderstanding*—This can be a result of undercommunication.
- *Interpreting*—Limited support for illustrating the overall relations between data might make the interpretation of data difficult.
- *Waiting*—If one depends on others filling in relevant data, this can result in unnecessary waiting.
- *Extra processing*—If wrong data has been provided, extra processing is necessary to correct it.

5.2.1 Knowledge Creation in Lean Engineering Environments

An essential aspect relative to the goal of minimizing waste is to be able to capture knowledge as it arises in the collaborative engineering activities, and distribute relevant knowledge to relevant people in a timely fashion; in-context and on relevant platforms. This is specifically relevant to address undercommunication, misunderstanding, interpreting, waiting (although enforcing data input without a specific plan for this might result in extra processing, and increase the searching problem).

Given the sources of waste identified and prioritized by the industry partners, LEAP could add value by supporting a continuum from informal, ad hoc collaboration (low-threshold mechanisms for capturing knowledge created as a result of ad hoc creativity in teams) to more formalized engineering processes.

A combination of pull- (search) and push-based knowledge provision mechanisms could represent added value in complex, multidisciplinary, and collaborative situations such as "knowledge assistants" where the person(s) in need of knowledge may or may not exactly know what to look for, how to retrieve it, or what data could be of interest.

5.2.2 Supporting Knowledge Creation in a Virtual Obeya

Whereas knowledge often arises in collaborative tasks on the individual and (small)-group level, and the iterative shifts between user collaboration and

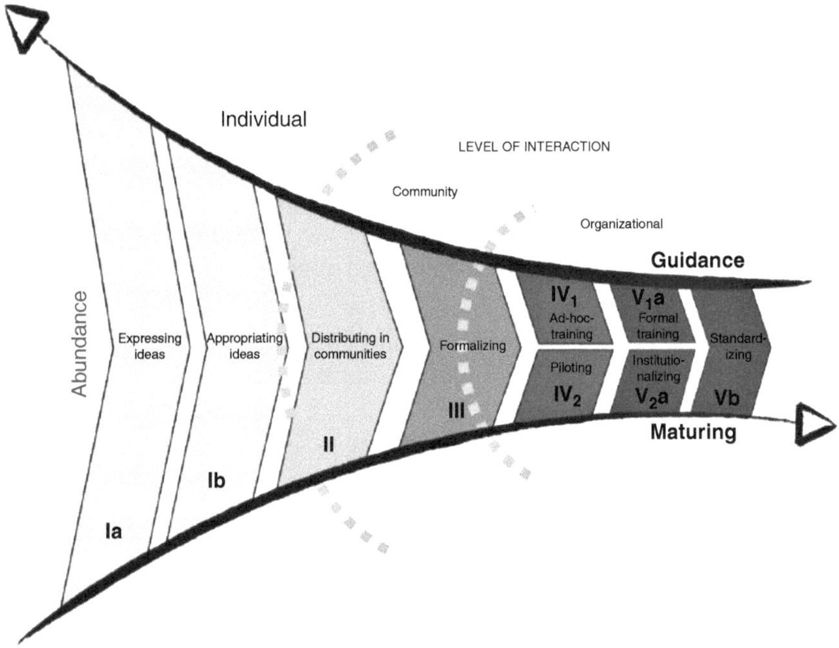

Figure 5.1 ***Framework for knowledge maturing.*** *(From Kump et al. [13]).*

automated design processes, it is in many cases also relevant to spread the knowledge established from this level to the organizational level. Note that it is not given that we want developed knowledge to be shared through all of the organization. To discuss this in more detail, we base the discussion on the work on knowledge maturing from the EU FP7 MATURE project. The following is based on Kump et al. [13].

The Knowledge Maturing Model outlines the following phases (see Fig. 5.1):

- *Ia. Expressing ideas*—New ideas are developed by individuals either in informal discussions or by 'browsing' the knowledge available within the organization and beyond. Extensive search and retrieval activities potentially result in loads of materials facilitating idea generation.
- *Ib. Appropriating ideas (individuation)*—New ideas that have been enriched, refined, or otherwise contextualized with respect to their use are now appropriated by the individual. Knowledge sources and new knowledge are "bookmarked" so that an individual can benefit from their future (re-)use.
- *II. Distributing in communities (community interaction)*—This phase is driven by social motives such as belonging to a preferred social group or the

expectation of reciprocal knowledge exchange within the community or project. A common terminology for individual contributions is developed and shared among community members.

- *III. Formalizing (information)*—Artifacts created in the preceding phases are often unstructured and restricted to a limited local context. They are only comprehensible for people in this community, as shared knowledge is still needed for interpretation. In Phase III, structured documents are created in which knowledge is made more transferable, and context is made explicit with the purpose to ease the transfer to people other than those in the originating community or project.

From Phase IV on, there are two alternative paths of knowledge maturing:

- *IV_1. Ad hoc training (instruction)*—Activities related to creating training materials out of documents that are typically not suited as learning material as they have not been made with didactical considerations in mind. Topics are refined to ease teaching, consumption, or reuse. Tests help assess the knowledge level and select learning objects or paths. Knowledge can be used for formal training in Phase *V ($V1_a$. Formal training (instruction)*. The subject area becomes teachable to novices. A curriculum integrates learning content into a sequence using didactical concepts to guide learners in their learning process. Learning modules and courses can be combined into programs used to prepare the student for taking on a new role, for example.
- *IV_2. Piloting (implementation)*—Experiences are deliberately collected with a test case stressing pragmatic action trying a solution before a larger roll-out of a product or service to an external community, or new rules, procedures, or processes to an internal target community, such as project teams or other organizational units. Know-how can be institutionalized at the beginning of Phase *V.*
- *$V2_a$. Institutionalizing (introduction)*—Within an organization, formalized documents that have been learned by knowledge workers are implemented into the organizational infrastructure in the form of business rules, processes, or standard operating procedures. In the organization-external case, products or services are launched on the market.
- *V_b. Standardizing (incorporation)*—This latest phase covers standardization or certification. Certificates confirm that participants of formal trainings achieved a certain degree of proficiency or justify compliance with a set of rules that the organizations have agreed to fulfill. Standards also help connecting products or services or showing that they fulfill laws or recommendations before being offered on a certain market.

Often, the individual engineer is working by himself or herself with his or her tools (eg, CAD tools) that are made to support specific tasks. When meeting for collaboration in the Virtual Obeya, knowledge from different experts is brought together and can be presented in an interrelated way, also taking other context information into account. In such sessions, in addition to learning from each other, one often also see that new knowledge is produced, and it is often important to be able to represent this for later use. This knowledge might be relevant on different levels (referring to the levels in the previous model):

1. For the workers only, to be used as private guidelines;
2. For the whole project;
3. For a wider community of experts within an area;
4. For the whole department or the whole organization/enterprise (different levels can be relevant here based on the size of the organization and the organizational structure); and
5. Beyond the organization (eg, if it is input to existing standards in the process of further developing the standard); other examples are if one is part of an extended enterprise or a network involved in open innovation.

Note that the learning and experience will often need to be more restricted the more widely that one wants to share the knowledge. Thus, whereas it is beneficial to capture knowledge at the source, new knowledge captured should have different visibility depending on its nature and the goals of knowledge creation.

Another dimension is the formality of the knowledge captured, which is very much related to the level of knowledge maturity that one wants to achieve, it can be:

1. Informal annotations;
2. Relations (eg, as openlinked data), to or between existing knowledge sources can be added;
3. Annotations linked to the context (thus potentially linked to concepts in the common ontology);
4. Codified knowledge (eg, as new formal KBE-rules);
5. Structured to act as basis for ad hoc or organizational training;
6. Structured to act as a tool, product, or process to be reused across the organization;
7. Structured to act as input to external standards.

The availability and formality of the captured data might also be temporally restrained (eg, all new knowledge is kept within the project until the end of the project) and first then one shares information more widely (eg, through reflection session) shared either on instance level (eg, as examples of good or bad practices) or on the type level (eg, updating some methodology used across the organization). In all cases, it is important to be clear on the level of knowledge maturing that you are on and the ambitions to increase the level of maturity. Aiming too high will result in waste due to extra processing. Aiming too low might result in waste through undercommunication, with possible result in waste due to unduly searching, interpretation work, and misunderstanding.

To effectively cope with complex data and information sources, context-driven approaches have been investigated in a research task on Semantic UI principles and context-based approaches for the Virtual Obeya. In the continuation of this work, initial research has been carried out in the area of context-aware recommender systems to further reduce waste and to improve knowledge access in multidisciplinary engineering projects.

5.3 CONTEXT-AWARE RECOMMENDER SYSTEMS AND KNOWLEDGE ACCESS IMPROVEMENT IN MULTIDISCIPLINARY ENGINEERING PROJECTS

During a project development process in a multidisciplinary work environment, there are challenges for engineers as knowledge workers while they are performing their assigned tasks; largely based on the large amounts of available information and the lack of tailored solutions for information seeking and access. Finding the information necessary for specific tasks is quite a time-consuming process and leads to lower development speed, as shown in [14] due to waste of searching, interpreting, undercommunicating, or waiting [15]. This challenge needs to be addressed.

Engineers of each discipline have different types of information needs according to their particular work situations. For example, they might need to find some facts or instructions, collect some advice for decision making, find examples to reuse, and so on. The way the users interact with a system for exploring their required information influences the management of knowledge resources [16].

To obtain the desired information, engineers often contact colleagues by emails or meetings to ask for the required knowledge, and also by searching through general or in-house search engines. Using search engines can help to accelerate knowledge access; however, engineers face some challenges, such as spending large amounts of time to gain relevant information and also finding the proper search keywords. This is especially true for engineers

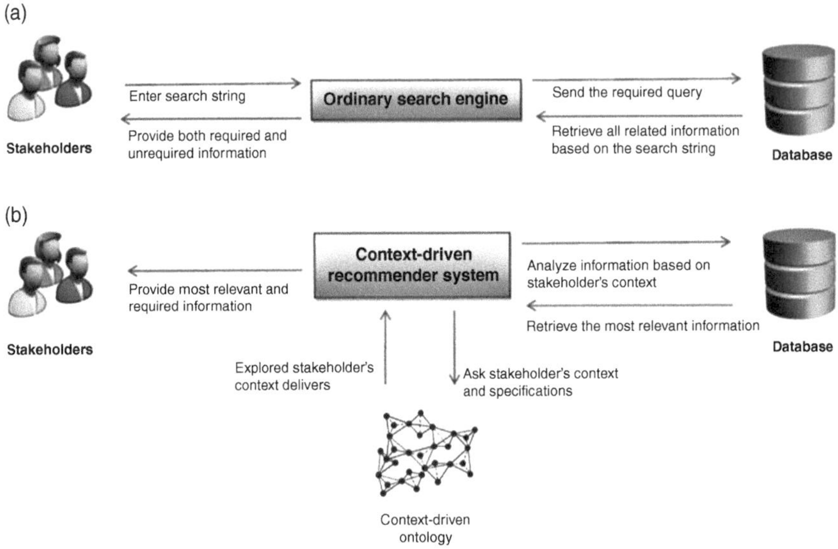

Figure 5.2 ***Engineers' interaction with information access systems.*** (a) Search engine, (b) context-driven recommender system [14].

who are new in the system or who have a lower level of expertise. Search results might not be semantically related to the particular situation of the engineer and an intelligent service is required to improve knowledge and information access by personalizing the information according to the engineers' situations in their working life. Such a system is known as a recommender system, which utilizes the context of the engineers and identifies those pieces of knowledge and information that match this context [14].

As illustrated in Fig. 5.2a, the engineers have to search through an ordinary search engine for information by entering a relevant search string and the search result returns any document that contains the search string without considering the engineering or business domain. Adding a semantic understanding of documents can improve this process. As shown in Fig. 5.2b applying a context-driven recommender system may provide more relevant results because it considers an engineer's particular work situation [14].

As a context-driven domain knowledge approach, the recommender system employs an ontology to add semantics to the process, for example in ontology-based information extraction systems [17], in order to match documents to an engineer's situation. Exerting a reference ontology for the whole domain to structure entities can help foster semantic annotation since entities are linked together by informative relationships. In addition to recommending relevant documents to the right engineers, recommending relevant people in the organization is also important to improve interaction in a design-centric engineering domain [18].

In addition, varying work situations of the users in different stages of the project should also be considered. For example, an engineer in a product design company can have overlapping social circles and implicit connections. One can be the organization chart of the hierarchy, setting her up in relation to people within her department, her supervisors, her team, or her staff. Additionally, she will be part of all the engineers that work in the company in different departments, and she may also be part of one or multiple projects. She might additionally be part of the user group for a specialized piece of software or be part of a management group. All these roles and tasks will demand different work and information access from her [18].

Note that a sole recommendation system with no user refinement can be insufficient. Recommendation should be adapted toward an improved navigation that presents suggestions, but is open for user refinement. Furthermore, there are different categories of recommendations based on different information needs. An option is to split the recommendations into different navigational categories from different perspectives [18]:

- hierarchically-related documents;
- related workflow tasks and workflows;
- similar projects and project overviews;
- similar and related parts of a similar project;
- similar type of document (project documents, module documentation, design drawings, lessons learned, best practices, etc.);
- different organizational perspectives (management, client relations, controlling, PR, ...).

There are some challenges for providing relevant recommendation for the target users. A major challenge is to properly understand and model both the relations of the users and their influence on the search task, and the content and relation of the documents. This is crucial in supplying the right information by direct recommendation, inform the ranking when she is searching for information, or help her better filter and manage data, documents, or knowledge objects [18]. Additionally, personalization goals can change a lot during a work task, as she can take on different roles. Finally, there is the problem of sparse and insufficient data for small-scale in-house systems because they contain a much smaller number of documents, users, and interactions than in large-scale systems, which can make statistical approaches biased or wholly inapplicable [19]. Using conventional approaches, this would put a recommender system into a permanent cold start condition.

By using a hybrid approach of both collaborative filtering and content analysis together with the ontology-based semantic approach, the impact of these issues can be reduced and a system even for this challenging environment can be developed [20].

5.4 TOWARD AN INTEGRATED COLLABORATION SYSTEM—MANAGING COLLABORATION FOR COMPETITIVE ADVANTAGE

Succeeding with collaboration is a complex undertaking, and few companies succeed in exploiting the full potential of deep collaboration. One of the main reasons for this is that collaboration is suffering from *major coherency disconnects*

First, improvement efforts often fail because strategic initiatives and decisions are not followed up by operational measures—instating a policy that collaboration constitutes a main element in running the business does not lead to change unless it is followed by specific, clearly defined work practices that spell out how to use collaboration operationally to achieve business objectives. Collaborative engineering must not remain a loosely defined, ambiguous term—it must be given a clearly defined content. As one example, shown in Chapter 7, the Virtual Obeya can be a powerful concept when it is positioned in a wider context—with a clear purpose, and supporting methods/practices and other elements of a wider approach where the different elements mutually reinforce each other.

Second, improvement efforts often fail because collaboration is seen and treated as something that is domain-specific rather than an enterprise-wide concept. Collaboration does not equal a single collaboration tool or platform, or even a set of tools or platforms; indeed most activities in engineering companies today include some collaborative aspect(s). Unless these misconceptions are cleared and a proper understanding of how to facilitate and manage collaboration as a broad set of business activities (often across a product lifecycle) is in place, it will be difficult to reap the full benefits of collaboration—and equally avoid misapplications of collaboration (that can reduce performance).

5.4.1 A Process for Improving End-User Collaboration

The value of collaboration in engineering contexts as they can be observed in LinkedDesign industry partners' respective organizations is strongly related to collaboration as a purposeful activity. Furthermore, collaboration is dynamic, it evolves over time and coexists in a large and growing number of different forms, where a range of different data, information, and knowledge sources are consulted individually or collaboratively, in the context of daily engineering activities.

Effectively supporting high-performance collaborative work patterns across a variety of different engineering contexts requires competencies for both individual knowledge workers and process owners. As described by Hansen and Nohria [5]:

> *Firms come into being in order to enable human beings to achieve collaboratively what they could not achieve alone. If one accepts this as the true purpose of any organization, then the main focus of executives' attention should be on how to foster collaboration within their companies.*

One could indeed argue that collaboration is the raison d'être for both teams and different organizations—enabling these teams or organizations to achieve different and better outcomes than they could have done as a disconnected group of individuals.

Moreover, performance is typically linked to collaboration on both a strategic and an operational level, both for work and reflection on work for learning and process improvement. While technology is an important enabler of new collaborative work forms with attractive characteristics, technology alone is not sufficient to enable new, high-performance lean engineering practices. Broader change initiatives including smart combinations of people resources, technology, work processes, business culture, and organizational models are needed to fully exploit the value of collaboration. This is further complicated by the current lack of well-known, industrial frameworks for: first, evaluating the impact of; and second, systematically improving collaboration.

There are, however, a number of tools and diagnostic frameworks that can be applied as decision support tools to make informed decisions that reduce risk and improve the chance of success, such as Hansen [21,22], Rosen [23] and Mattesich et al. [24]—additional information can be found in Table 5.1.

As a continuation of this work, an initial collaborative diagnostics toolbox has been compiled, containing a repository of decision support tools that can be used, such as when addressing commonly occurring sources of waste in collaboration. When exploring and piloting in-context collaborative work in environments developed for effective and efficient knowledge sharing, visualization and joint decision making, these tools can be used to add value by accurately pinpointing nontechnology collaboration barriers and appropriate solutions, improving overall performance.

Table 5.1 Process template for improving user collaboration

Objective/process step	Description
Promote awareness	Finding relevant examples from a range of industries to identify a suitable level of ambition
Define precise collaboration objectives	Using available knowledge to define a set of precise, actionable objectives that links collaboration to overall business objectives
Identify collaboration barriers	Using state of the art diagnostics tools to identify and benchmark collaboration barriers
Define precise collaboration requirements	Using a comprehensive database as an external resource when working systematically with eliciting user requirements
Additional analyses and success factors	Additional fit-for-purpose analysis; overall collaboration system and collaboration success factors
Identify solutions	Decision support with respect to identifying solutions that fit the problem, whatever the problem may be (process, organization, technology, culture)
Realize benefits	Decision support with regards to ensure learning is put to use through alignment of chosen solution(s), smooth operations and continuous improvement

5.4.2 Managing Human Success Factors in Collaboration

Human factor robustness is achieved through applying open, transparent leadership and making everybody part of the team. Tips include stating explicit, unifying goals, speaking a language of collaboration, and managing collaboration success factors.

- In its simplest form, undertake simple investigations about your performance according to a defined framework built around some core principles or dimensions and then work systematically to improve your performance along those dimensions. Examples of simple frameworks that could support light initiatives include Deloitte's "As One" framework built around the three core conditions Belong, Believe and Behave [25], or ten commonly occurring sources of waste in collaboration [10].
- For a more thorough analysis with using validated tools with benchmark data, Hansen's methodology to create and sustain disciplined collaboration patterns [21], see also [5], is a good alternative for those interested in achieving better results through collaboration. To assess frequently

occurring collaboration barriers, a built-for-purpose survey from Morten T. Hansen's methodology "The Collaboration Toolkit" [22] can be used. This methodology enables benchmarking against a sample of $n = 107$ companies across various industries, and also contains many other diagnostics tools.

- Similarly, Mattessich et al. [24] in their milestone study "Collaboration: What makes it work" identified 20 success factors for collaboration based on a very comprehensive literature review of 40 studies selected on the basis of validity and relevance from a total pool of 414 studies describing success factors in the field of collaboration. The 20 success factors are grouped into the six categories: environment, membership characteristics, process and structure, communication, purpose, and resources. This research has been used to develop the Wilder Collaboration Factors Inventory [26]—a useful tool that makes it possible to assess how your collaboration efforts score on 20 research-tested success factors. You will be able to view factor scores as well as item averages and open-ended responses for your group's completed inventories.

For a comprehensive approach, Appendix A contains a repository of tools and activities for the objectives/process steps outlined in Table 5.1, which can be used as decision support when working to systematically improve user collaboration, and when addressing commonly occurring sources of waste in collaboration.

When exploring and piloting in-context collaborative work in environments developed for effective and efficient knowledge sharing, visualization and joint decision making, these tools can be used to add value by accurately pinpointing nontechnology collaboration barriers and appropriate solutions, improving overall performance.

5.5 CONCLUSIONS

The LEAP collaboration system is a compilation of models, concepts, elements, and technology components that—when combined and structured in a meaningful way to teams of end users—enable companies to execute split-location engineering projects in a way that represents a competitive advantage.

This chapter with the related appendix provides some perspectives on a systematic, structured approach to enabling and facilitating effective and efficient collaboration in engineering teams, that can complement, enrich and increase the impact of some of the LEAP tools and services, including the Virtual Obeya.

The realization of expected benefits requires a clear collaborative strategy, and daily, hands-on operations. Lack of either (or a clear connection between the two) dramatically reduces the success rate (for additional details, see eg, [27] and [28]).

Written specifically for split-location and multidisciplinary teams, this chapter contains a description of collaboration elements in this context. Furthermore, this chapter describes how to develop and operationalize such a system in a way that it represents a holistic, meaningful, and value-adding collaborative working environment that enables engineers and other knowledge workers to make decisions, solve problems, and address multidisciplinary issues effectively and efficiently.

Based on lean thinking, the LEAP collaboration system approach identifies ways of reducing waste in collaboration processes. An evaluation of knowledge sources is described, together with approaches supporting knowledge creation in lean engineering environments. A unified approach and a diagnostics toolbox have been developed to help companies to systematically improve collaboration in their teams.

ACKNOWLEDGMENT

The majority of this chapter is based on the following project report: Bakås O, Krogstie J, Sjøbakk B, Petersen SA, Stokke F, Sriram P, et al. D5.2 Lean engineering collaboration system diagnostics: optimization handbook & workbench. LinkedDesign project report, LinkedDesign, FoF-ICT-2011.7.4 Project No: 284613; 2013. Some sections have been updated with research results from the final phase of the LinkedDesign project.

REFERENCES

[1] Williams B, Figueiredo J, Trevelyan J. Engineering practice in a global context: understanding the technical and the social. Leiden: CRC Press; 2013.
[2] Butler P, Hall TW, Hanna AM, Mendonca L, Auguste B, Manyika J, et al. A revolution in interaction. McKinsey Q 1997;February(1).
[3] Beardsley SC, Johnson BC, Manyika JM. Competitive advantage from better interactions. McKinsey Q 2006;May(2).
[4] Morello D, Burton B. Future worker 2015 extreme individualization. Gartner Research Report, Gartner Research; 2006.
[5] Hansen MT, Nohria N. How to build collaborative advantage. MIT Sloan Manage Rev 2004;46(1):22–30. World Economic Forum, Davos, Switzerland, January 2006.
[6] La Rocca G. Knowledge based engineering: Between AI and CAD. Review of a language based technology to support engineering design. Adv Eng Inform 2012;26(2):159–79.
[7] Hermans FFJ. Analyzing and Visualizing Spreadsheets. PhD Thesis. Software Engineering Research Group, Delft University of Technology, The Netherlands; 2012.

[8] Blackstone JH. APICS dictionary. Athens, GA: APICS; 2008.
[9] Kelle P, Akbulut A. The role of ERP tools in supply chain information sharing, cooperation, and cost optimization. Int J Product Econ 2005;93–94:41–52.
[10] Manyika J, Sprague K, Yee L. Using technology to improve workforce collaboration. What Matters. McKinsey Digital; 2009.
[11] Lyons MH. Future ICT systems—understanding the business drivers". BT Technol J 2005;23(5):11–23.
[12] Grudin J. Groupware and social dynamics: eight challenges for developers. Commun ACM 1994;37(1):92–105.
[13] Kump B, Knipfer K, Pammer V, Schmidt A, Maiier R, Kunzmann C, Cress U, Lindstaedt S. The role of reflection in maturing organizational know-how. In: Proceedings of 6th European conference on technology enhanced learning (ECTEL); 2011.
[14] Mehrpoor M, Gjaerde A, Sivertsen OI. Intelligent services: a semantic recommender system for knowledge representation in industry. Proceedings of the 20th ICE & IEEE-ITMC conference on engineering, technology and innovation. Bergamo, Italy; 2014.
[15] Kristensen K, Aasland KE, Blankenburg D, Sivertsen OI. Virtual Obeya: collaborative tools and approaches addressing sources of waste in lean engineering. Proceedings of the 19th ICE & IEEE-ITMC international conference responsible innovation and entrepreneurship. The Hague, the Netherlands; 2013.
[16] Freund LS. Exploiting task-document relations in support of information retrieval in the workplace. PhD Thesis. University of Toronto; 2008.
[17] Wimalasuriya DC, Dou D. Ontology-based information extraction: an introduction and a survey of current approaches. J Inform Sci 2010;39:211–24.
[18] Ahlers D, Mehrpoor M. Semantic social recommendations in knowledge based engineering. SP 2014: Workshop on Social Personalization at Hypertext 2014, CEUR-WS. vol. 1210; 2014.
[19] Tait J. Issues and non-issues in professional search. Integrating IR technologies for Professional Search Workshop at ECIR2013; 2013.
[20] Mehrpoor M, Gulla JA, Ahlers D, Kristensen K, Sivertsen OI, Ghodrat S. Using process ontologies to contextualize recommender systems in engineering projects for knowledge access improvement. European conference on knowledge management (ECKM); 2015.
[21] Hansen MT. Collaboration: how leaders avoid the traps, create unity, and reap big results. Boston: Harvard Business Press; 2009.
[22] Hansen MT. The collaboration toolkit. Cambridge, MA: Harvard Business Press; 2009.
[23] Rosen E. The culture of collaboration. San Francisco: Red Ape Publishing; 2007.
[24] Mattessich PW, Murray-Close M, Monsey BR. Collaboration: what makes it work. second ed. St. Paul, MN: Fieldstone Alliance; 2001.
[25] Frederick D, Miller DB, Andrew G. As one—better collaboration where it counts the most. Deloitte Rev 2013;12.
[26] Wilder Collaboration Factors Inventory (online collaboration assessment). http://wilderresearch.org/tools/cfi/; 2015.
[27] Kristensen K. Collaborare non humanum est: Collaboration et environnements hybrids (French translation by Françoise et Loukhoum Bronner). Office et Culture Magazine, June ed.; 2010, pp. 32–39.
[28] Kristensen K, Kijl B. Collaborative performance: addressing the ROI of collaboration. Int J eCollab 2010;6(1).
[29] ECOSPACE (eProfessional Collaboration Space) project website, project funded under IST-2005-2.5.9. Collaborative working environments. http://cordis.europa.eu/project/rcn/80736_en.html; 2015.
[30] Frost & Sullivan. Meetings around the World II: charting the course of advanced collaboration. Frost & Sullivan Whitepaper; 2009.

APPENDIX A: COLLABORATIVE DIAGNOSTICS: TOOLS FOR LEAN ENGINEERING COLLABORATION SYSTEM ANALYSIS AND OPTIMIZATION

The following table contains a repository of tools and activities that can be used as decision support when working to systematically improve user collaboration, and when addressing commonly occurring sources of waste in collaboration.

When exploring and piloting in-context collaborative work in environments developed for effective and efficient knowledge sharing, visualization, and joint decision making, these tools can be used to add value by accurately pinpointing non-technology collaboration barriers and appropriate solutions, improvingw overall performance.

Table A.1 Tools for lean engineering collaboration system analysis and optimization

Objective/ process step	Tools, literature, and decision making support
Awareness	*Tools, literature and processes:* Literature—What results have others achieved through improved collaboration, based on the same or similar strategies? *Results/typical outcome:* Statistics—Qualified statistics and results that can be used for motivational purposes and for setting the right ambition level with respect to achievable, verifiable and (in many cases) quantifiable results. Case studies; what others have done (specific approaches) and achieved (results). Rationale—In-depth understanding of how collaboration impacts different aspects of business, such as innovation, sales and operational efficiency.
Objectives	*Tools, literature, and processes:* Upside calibration/benchmarking tool for multiunit companies, potential benefits from improved collaboration in the key business areas cross-unit innovation, cross-unit sales, and cross-unit operations [22]. Benchmarking tool against a sample of 107 companies, including potential — current status gap analysis. The collaboration matrix; allows you to pinpoint collaboration opportunities between two units. Pairwise comparison; the benefit is that the assessment becomes more granular as every possible pair of units is scrutinized. Upon completion of the assessment, you can then add up all possibilities to view the grand total [22].

Table A.1 Tools for lean engineering collaboration system analysis and optimization (*cont.*)

Objective/ process step	Tools, literature, and decision making support
	Literature—What is the link between existing strategies and specific business objectives and the specific/potential contribution of collaboration in fulfilling these strategies and objectives? *Results/typical outcome:* Collaboration objectives—Understanding of specific benefits of improved cross-unit collaboration as a basis for precise definition of objectives. Derive and develop collaboration objectives from strategic business objectives and policy documents. Collaboration processes—Define target collaboration processes for improvement, linked to collaboration objectives. Metrics—Define collaboration metrics. Establish baseline for comparison based on results from comparable cases.
Barriers	*Tools, literature, and processes:* The Four-Barrier Rating—defines the score of four commonly observed collaboration barriers, as a basis for corrective action and lean collaborative performance [22]. Benchmarking tool against a sample of 107 companies. Sources of waste in collaborative processes assessment, based on Manyika et al. [10]. *Results/typical outcome:* Barrier identification and benchmarking—Identify the most important barriers to collaboration, for benchmarking and prioritization of actions with high precision level.
Requirements	*Tools, literature, and processes:* Literature—Miscellaneous requirements, case studies and comparable data from numerous roadmapping projects, industry projects and other research projects such as ECOSPACE [29]. *Results/typical outcome:* Requirement elicitation—In-depth analysis of multistakeholder requirements, including end users and support personnel. Based on European research projects. Requirement prioritization—Review of requirements and priority setting. User interface requirements—Establish activity-centric user interface principles.
Analysis	*Tools, literature, and processes:* Depending on the context, analysis tools include, but are not limited to the following: The barrier-to-lever worksheet [22]—Different barriers require different solutions. This tool allows you to assess which levers need to be deployed in your situation, given the set of barriers you are confronting. Deals mostly with HR, organizational and cultural aspects.

(*Continued*)

Table A.1 Tools for lean engineering collaboration system analysis and optimization (*cont.*)

Objective/ process step	Tools, literature, and decision making support
	Success factor scorecard—Wilder Collaboration Factors Inventory; most important success factors for collaboration identified from 40 (18 + 22) out a total of 414 (133 + 281) studies. An online collaboration assessment based on commonly observed collaboration success factors [26]. Scenario analysis—Different "overlays" highlighting different aspects of comparable scenarios. Priority collaboration processes analysis—Simple business process criticality and frequency matrix. Business needs—behaviors—building characteristics mapping and matching tool. Current-future gap analysis—including identification of enablers Collaboration capability and maturity assessment—based on a comprehensive list of collaboration technologies. Collaboration stakeholder analysis—based on RASCI analysis. Collaboration premium assessment—used to qualify typical processes and their collaborative potential [22]. *Results/typical outcome:* Fit-for-purpose levers—for subsequent consideration. ROI (Return on Investment)—and business case preparation and analysis. ROC (Return on Collaboration) analysis—where this is more appropriate than an ROI analysis [30].
Solutions	*Tools, literature, and processes:* Literature—miscellaneous solutions, concepts, success factors, case studies and comparable data from numerous roadmapping projects, industry projects and research projects. The Collaboration Lever Barometer [22]—used to assess the status of collaborative levers. Benchmarked data. The Collaboration Lever Gap Analysis [22]—analyswes gap between desired and current use. Collaborative Leadership Style and Benchmarking Tool [22] (including profile and personal barrier diagnosis)—to analyze and develop productive, collaborative behaviors. *Results/typical outcome:* Development of collaboration concepts, enablers, levers and technologies—linked to barriers and success factors. Total collaboration concepts.

Table A.1 Tools for lean engineering collaboration system analysis and optimization (*cont.*)

Objective/ process step	Tools, literature, and decision making support
Benefits	*Tools, literature, and processes:* Literature—miscellaneous solutions, concepts, success factors, case studies and comparable data from numerous roadmapping projects, industry projects and research projects. Collaboration system analysis and optimization—"due diligence" of collaboration enablers and inhibitors in different categories. Feedback and learning *Results/typical outcome:* Proper context-specific interpretation of results—activated tools and processes listed earlier. Training activities—how to tweak the implemented solution for maximum benefit. Coaching and advisory activities—how to develop feedback loops and continuous improvement principles for lean collaborative operations. Collaboration leadership and management principles—for moving forward. Feedback and lessons learned.

CHAPTER 6

LEAP Interoperability Standards

K. Främling*, A. Buda*, S. Kubler*, J. Cassina, E. Coscia**, S. Parrotta**, S. Terzi† and D. Cerri†**
*Aalto University, Espoo, Finland
**HOLONIX S.R.L., Meda, Italy
†Polytechnic of Milan, Milan, Italy

Contents

6.1 BACKGROUND

A major obstacle to effective, whole-of-life lifecycle management is that valuable information is locked into vertical applications, sometimes called "silos." This information is not readily accessible for other interested parties across the Beginning-of-Life (BOL), Middle-of-Life (MOL), and End-of-Life (EOL) lifecycle phases. Closed-Loop Lifecycle Management (CL2M) [1] is a concept that has been developed to describe the accessibility and use of information between all relevant organizations and information systems throughout these phases. CL2M signifies a major leap beyond "traditional" Product Lifecycle Management (PLM). CL2M also provides a framework that contributes to methods such as Life Cycle Cost (LCC) analysis and Lifecycle Cost Assessment (LCA).

Common, open, and trustworthy interoperability standards for CL2M will enable the closing of information loops, allowing information to be shared across the whole spectrum of lifecycles, such as product, human, food and beverage, pharmaceutical, healthcare, supply chain and logistics, and

Taking the LEAP
http://dx.doi.org/10.1016/B978-0-12-805263-1.00006-2

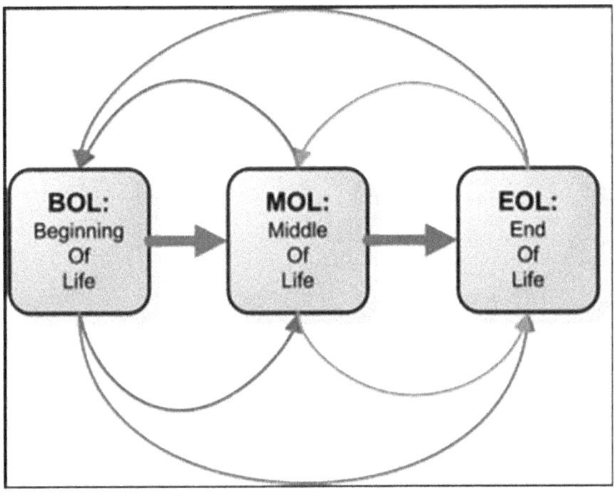

Figure 6.1 ***Closing information loops*** [2].

data governance, among many others. A major objective of the LEAP developed in LinkedDesign is to be able to handle all relevant CL2M information so that information from any single lifecycle phase may affect processes and decision making in the other phases. For example, information about the condition of products at EOL may be fed back and used to affect the maintenance of similar products during MOL or to improve the design and production of future product series at the BOL. Closed loops ensure that valuable information is available to all lifecycle phases (Fig. 6.1).

According to the currently dominant definition of the Internet of Things (IoT) [3], it provides the necessary tools for building information systems that enable CL2M. The "Things" are rapidly becoming more and more equipped with embedded sensors, processing power, and communication capabilities. Such increasingly self- and environment-aware and communicating entities have also been called Intelligent Products in the literature [4,5]. Intelligent Products have the means to interoperate between each other and with other information systems. Intelligent Products also play an essential role in CL2M because they are able to collect usage information and react on it proactively, such as estimating needs for maintenance or repair [6]. To let Intelligent Products interoperate with each other and with other information systems, interoperability standards are needed. In the LinkedDesign project, such interoperability is also needed between the different software components that are used for building the LEAP. This chapter presents two interoperability standards developed for this purpose, which are published by The Open Group: the Open Messaging Interface (O-MI) [7] and the Open Data Format (O-DF) [8]. This chapter describes the design principles and provides a description of the standards, including implementation principles and examples of real-life implementations.

6.2 INTERNET OF THINGS STANDARDS

PLM is commonly understood to be a strategic approach that incorporates the management of data associated with products of a particular type, as well as the business processes that surround it [9]. These product definition data typically consist of detailed specifications, user manuals, CAD drawings, manufacturing instructions, service manuals, disposal, and recycling instructions and so forth. Interorganizational data exchange is a prerequisite for any kind of PLM in the context of the extended enterprise [10]. In PLM, it usually signifies the transfer of electronic documents or business data from one computer system to another, that is, from one trading partner to another without human intervention. Data exchange requires some kind of physical media for transmitting the data from one system to the other, a protocol that allows us to determine what a system should send, when and where, one or more interfaces that send and/or receive data, and a semantic layer for understanding the data in a uniform way. The further up we go on the "protocol stack" towards the semantic level, the more challenging it is to create data exchange standards that are unambiguous while providing a sufficient power of expression. When dealing with product instance-enabled PLM, the need for adequate standards becomes even more essential due to the presence of people, embedded devices, and mobile devices that need to communicate with each other in ad hoc and context-dependent ways.

This product instance-enabled PLM was the cornerstone of the information systems developed in the PROMISE EU FP6 project (http://promise-innovation.com/). Real-life industrial applications requires the collection and management of product instance-level information for many domains involving heavy and personal vehicles, household equipment, phone switches, and so on. Information such as sensor readings, alarms, assembly, disassembly, shipping event, and other information related to the entire product lifecycle needs to be exchanged between several organizations. The project consortium investigated potential messaging standards that would provide the kind of interorganizational information exchange needed for the IoT. When looking for suitable candidates that would fill the application requirements, some entire families of standards were eliminated. One such family of standards are the "low level" machine-to-machine (M2M) standards that are usually specified for Intranet environments. M2M protocols and standards also tend to be developed for very specific purposes and domains, which makes them unsuitable for use as generic application-level IoT messaging standards. An extensive study of those protocols and standards is out of the scope of this chapter but a detailed comparison study can be found in Ref. [11]. As a result, PROMISE created two main specifications: the PROMISE Messaging Interface (PMI) and the PROMISE System Object Model (SOM).

At the end of the PROMISE project, the work on these standards proposals was moved to the IoT Work Group of The Open Group (http://www.opengroup.org/getinvolved/workgroups/iot). PMI was later split into two standards: the Open Messaging Interface (O-MI) and the Open Data Format (O-DF), which were published as official standards of The Open Group in Oct. 2014. SOM has been modified and renamed accordingly and is now called Open Lifecycle Management (O-LM) that is being developed as a future standard of The Open Group. The IoT Work Group's goal is to identify relevant standards for enabling CL2M, as well as identifying potential gaps and to create standards that fill those gaps.

6.2.1 Open Data Format O-DF

O-DF is specified as a simple ontology using XML Schema, which is generic enough for representing "any" object and information that is needed for information exchange in the IoT. It is intentionally defined in a similar way as data structures in object-oriented programming. It is structured as a hierarchy with an "Objects" element as its top element. The "Objects" element can contain any number of "Object" subelements. Fig. 6.2 gives

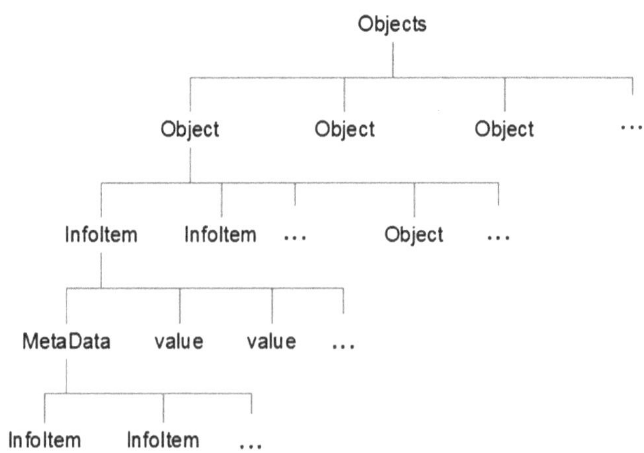

Figure 6.2 *Generic O-DF object tree.*

```
<?xml version="1.0" encoding="UTF-8"?>
<omi:omiEnvelope xmlns:omi="omi.xsd" version="1.0" ttl="-1">
    <omi:write msgformat="odf">
        <omi:msg xmlns="odf.xsd">
            <Objects>
                <Object>
                    <id>SmartFridge22334411</id>
                    <InfoItem name="FridgeTemperatureSetpoint">
                        <value>3.5</value>
                    </InfoItem>
                    <InfoItem name="FreezerTemperatureSetpoint">
                        <value>-20.0</value>
                    </InfoItem>
                </Object>
            </Objects>
        </omi:msg>
    </omi:write>
</omi:omiEnvelope>
```

O-MI

O-DF

O-MI

Figure 6.3 ***Example of a "write" message using O-MI and O-DF.***

insight into both the generic hierarchy/object tree while Fig. 6.3 shows an example of an O-MI message with an O-DF payload whose structure respects this object tree. In this example, a unique object of type Fridge (see line 7 of the XML message) is considered. "Object" elements can have any number of properties, referred to as InfoItems, as well as "Object" subelements. In our example, the Object Fridge has two InfoItems, named Door status and Thermostat (see lines 8 and 11 in Fig. 6.3). The resulting Object tree can contain any number of levels. Every Object has a compulsory subelement called "id" that identifies the Object (see line 7 in Fig. 6.3). The "id" should preferably be globally unique or at least unique for the specific application, domain, or network of the involved organizations.

In object-oriented programming, objects are aware of each other both by object containment hierarchies and by reference or pointers. In O-DF, such object references are made by using the Object "id" element. However, in the IoT, the "id" does not refer to a specific memory location but to an IoT object whose information may even be spread over several information systems and organizations. Different methods and systems have been proposed for the discovery of such distributed information [12]. The simplest mechanism is to include a URL in the identifier itself, as proposed by Ref. [13], and then to retrieve the information by object linking, as proposed by Ref. [14].

6.2.2 Open Messaging Interface O-MI

A defining characteristic of O-MI is that O-MI nodes may act both as a "server" and as a "client," and therefore communicate directly with each other or with back-end servers in a peer-to-peer (P2P) manner. Typical examples of exchanged data are sensor readings, lifecycle events, requests for historical data, notifications, and so on. O-MI messages are "protocol agnostic," so they can be exchanged using HTTP, SOAP, SMTP, FTP, or similar protocols. The most appropriate protocol to use depends on the application and the security mechanisms of the protocols. This nondependency on specific communication protocols makes O-MI different from many (or most) other potential IoT standards. Second, three basic but fundamental operations are defined in O-MI.

Fig. 6.4 illustrates these operations and variants. *O-MI write* is a basic update operation to modify one or several InfoItem values.

Fig. 6.4 illustrates such an operation, in which:

- A node denoted by @1 updates InfoItem d at $t = 1$ on a node denoted by @2, which then returns a success response.

Similarly, an *Immediate read* of information is a basic *O-MI read* operation to retrieve one or several InfoItems, as illustrated in

Fig. 6.4, in which:

- Node @1 retrieved InfoItem d on node @2 at $t = 1$, denoted by $d_{t=1}$.

The *Deferred read* of information, also referred to as subscription mechanism, is a cornerstone of O-MI. As illustrated in

Fig. 6.4, two types of subscriptions are available:

- *With callback address*: @1 subscribes to InfoItem d to @2, providing its address as callback with an interval of 2s.[a] Then, every 2s, both the current InfoItem value d and the subscription ID is sent by @2 to @1, that is, at $t = 1, 3, 5$.
- *Without callback address*: @1 subscribes to InfoItem d to @2 with an interval of 2 s but without providing its address as callback (denoted by null). @1 first receives a response with the ID of the subscription, and @2 memorizes d values every 2 s. Thus, @1 may request at any time one or several historical values with a read request including the subscription ID, as shown in Fig. 6.4, where @1 requests for values memorized at $t = 1$ and 5 (denoted by "Read(ID,$t = \{1, 5\}$)").

To *cancel* an existing subscription, node @1 simply needs to send a *Cancel* message with the ID of the subscription to node @2 (Fig. 6.4).

Note that the conceptual framework used for supporting subscriptions is the Observer Design Pattern, as presented by Gamma et al. [15], which signifies that an O-MI node can add itself as an observer of events that occur at another O-MI node. In this sense, O-MI differs from, for example,

[a]The corresponding function is noted "Subscription(d,x,add)" in Fig. 6.4; d the InfoItem to subscribe, x the interval value (in seconds), and the callback address.

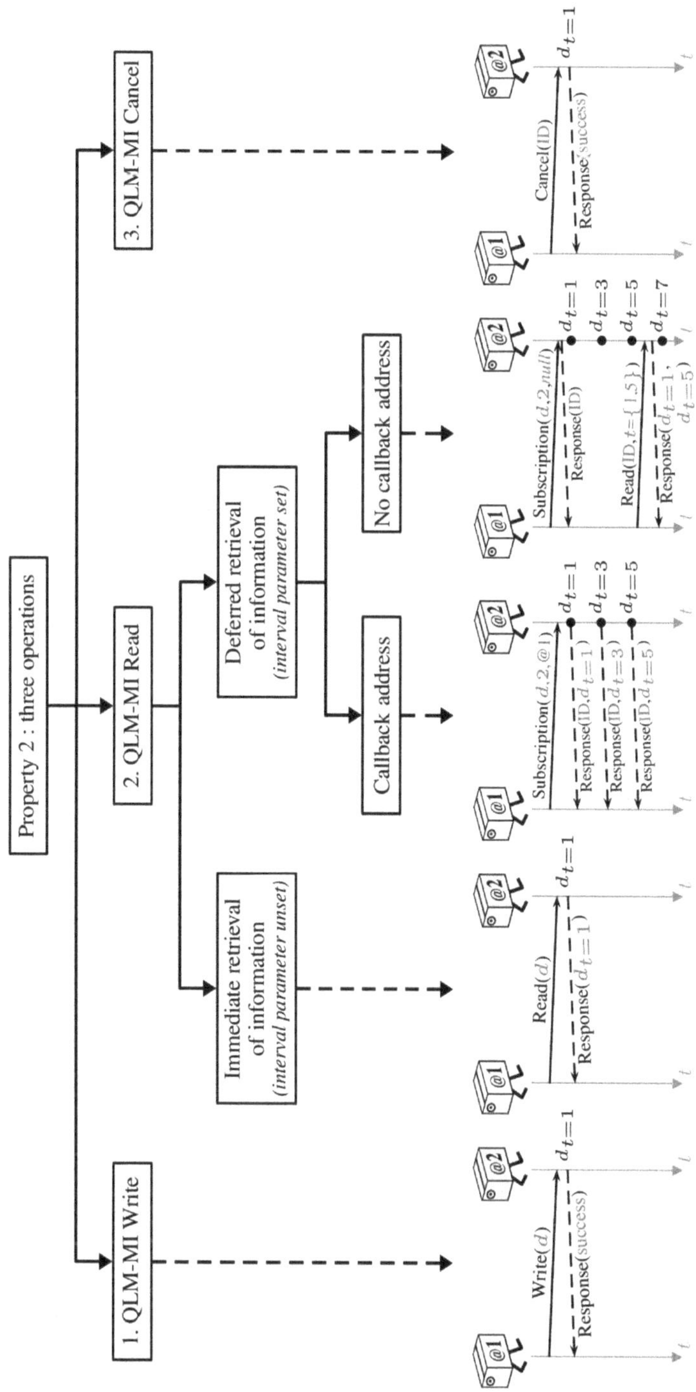

Figure 6.4 ***Basic operations and their variants defined in O-MI.***

JMS, which is based on the Publish/Subscribe (Pub/Sub) model. For many applications, the Observer and the Pub/Sub-models can be used in quite similar ways. However, the Pub/Sub-model usually assumes the use of a "high-availability server," which the Observer pattern does not. This is why the Observer model is more suitable for IoT applications where products might need to communicate with each other directly.

Another important feature of O-MI is that O-MI messages are "self-contained" in the sense that all the necessary information to enable the recipient to handle the message is contained in the message itself (eg, the actions to be performed, the callback address and so on). The O-MI message given as example in Fig. 6.3 highlights the message interface in which the TTL value is specified in row 2 and the operation type in row 3 (it is a "write" operation in this example).

6.3 LIFECYCLE MANAGEMENT STANDARDS FOR THE INTERNET OF THINGS

In the manufacturing domain, as in many others, there is a well-established shifting from product-centric to service-centric offers, where consumers will buy for the use of a product and for associated services rather than owning the product itself. This trend is dramatically affecting and transforming the value proposition and the value chain of the industry, creating the need to track each single product (or "physical item") along its whole lifecycle and to monitor the use experience. Existing IoT technologies can thus be combined with lifecycle knowledge management solutions to collect and use the large amount of data that can be collected to support the (re)design, not only of the product but also of the services, to predict and improve maintenance, and to support recycling and remanufacturing phases.

This information is generated, provided, and used by different systems and nodes in a distributed scenario, each one using different formats to represent them. Standards are, therefore, requested to remove the interoperability barriers and to enable the advertising, discovery, access, and exchange of the lifecycle information among these "nodes."

An analysis of the most relevant industrial standards in the field of product lifecycle and process data modeling has been conducted to check if and to which extent the aforementioned approach can be supported. The most relevant standards that were analyzed are:

- STEP (ISO 10303)
- STEP NC (ISO 10303-238, also known as AP238)

- PLCS (ISO 10303-239, also known as AP239)
- MANDATE (ISO 15531)
- PLM XML
- ANSI/ISA-95 (IEC 62264-1)

These standards have some properties and features in common, but they are also distinguishable by their many remarkable differences. First, they were designed by different organizations, with different scopes, and for different targets. STEP, STEP NC, PLCS, and MANDATE can at first sight be grouped together because they are all ISO (International Organization for Standardization) standards; furthermore, PLCS is an application protocol of STEP. To provide a process evaluation, several different process standards have been analyzed, some of these are aimed at representing a limited phase of a lifecycle or a process from a specific area (eg, MIMOSA), whereas others are more generally used to describe a wide variety of processes and activities. The ANSI/ISA-95 (ISO 62264) standard presented before can also be used to represent processes. The most relevant process standards that were analyzed are:

- ISO 18629—PLS (Process Specification Language)
- ISO 14224
- ISO 15926
- ISO/IEC TR 24774
- MIMOSA

See Fig. 6.5.

This analysis has pointed out that, so far, none of these standards has been conceived to deal with the information that are specific of a physical "intelligent" item; that is, of an instance of a product that, thanks to sensors an actuators is able to transmit information about its status and the surrounding environment, during its whole life.

The O-LM vision is to develop a standard that merges together product and process information that are collected from sensors and exchanged by connected items across all the phases of the product lifecycle. In particular, O-LM allows us to cover product information and related processes during the BOL, including specifications documents during the design, material and components used during the production as well as related processes with particular focus on MOL and EOL through sensors data retrieving and tracking the activities occurred during the usage and the disposal phases.

6.3.1 Open Lifecycle Management O-LM

O-LM is an extension of the O-DF standard that is used to exchange information regarding physical things characterized by a univocal identification

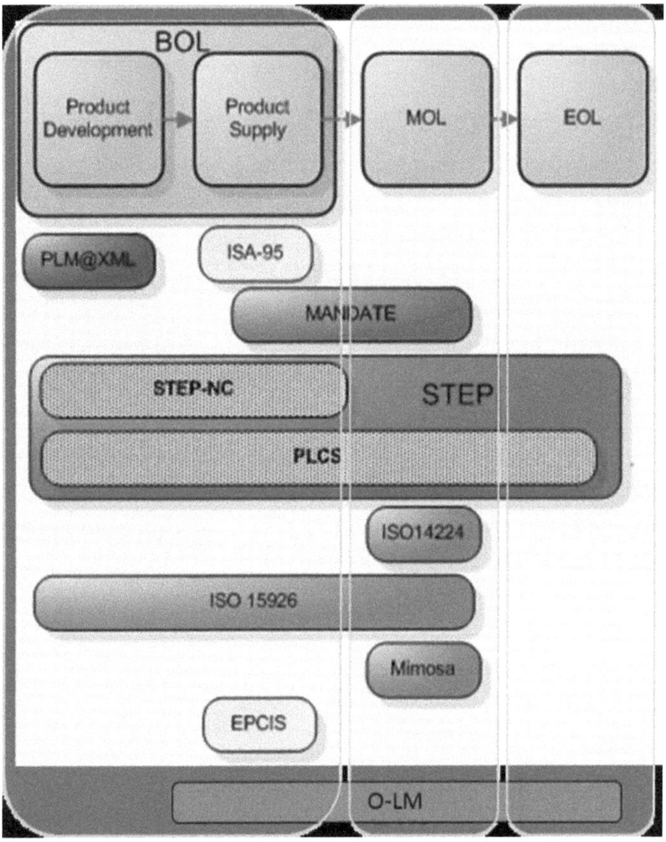

Figure 6.5 ***Most common product and process data standards.***

code. Whereas O-DF describes object information using a simple and generic format, O-LM focuses on products rather than on generic objects and on associating the Lifecycle dimension to the data describing the product.

Moreover, especially when dealing with advanced product/services solutions, the information that needs to be analyzed belong to physical products that are part of the IoT, thus may be equipped with sensors that collect and transmit different types of data to the Internet

As said previously, O-LM is an extension of the O-DF used to specify Physical Product information. The two models use a similar hierarchical structure. The common elements between the Object schema (O-DF) and Physical Product one (O-LM) as shown in Fig. 6.6 are:

- *Id* is used to identify Objects and Physical Products elements, the Id should be globally unique or at least unique for the specific applications;
- *Description* provides an human readable description;

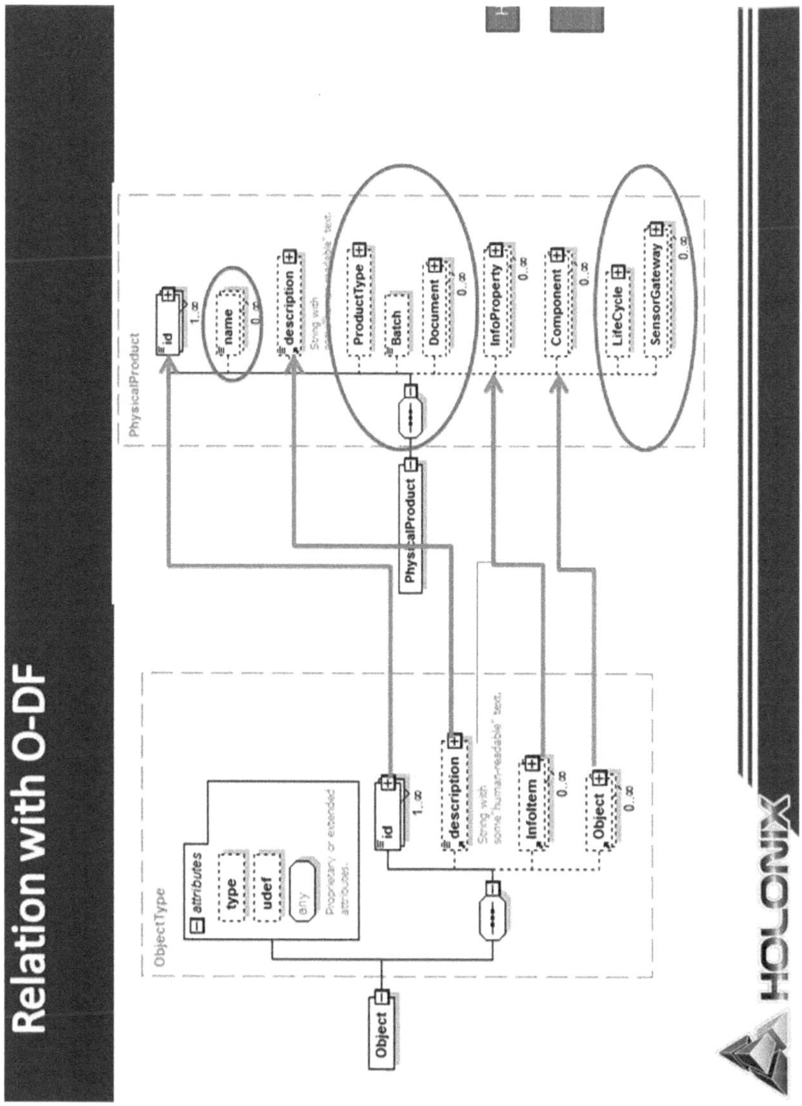

Figure 6.6 *Relations between concepts in O-DF and O-LM.*

- *Object* and *Component* elements are used to include further Objects and Physical Products within the main element;
- *Info Item* and *Info Property* are used to include an arbitrary number of subelements within Object and Physical Product classes that are not part of the elements already described within each model.

Within O-LM, the Physical Product element uses additional information defined through ad hoc fields:

- The *Name* is used to define the name of the Physical Product, it is a simple element represented by a string;
- The *Batch* element is used to define the manufacturing batch number of the Physical Product, it is a simple element represented by a string;
- The *Product Type* is a complex element used to describe the type of the Physical Product element and eventually relations with other Product Types;
- The *Document* element is used to describe Physical Product specifications, there could be several types of documents, such as CAx, text, video and so on, with different versions.
- The *LifeCycle* element is used to record all processes occurred during the product lifecycle. The element shows what happens during a specific moment of the product life, where this event occurs, and if there are relations with external resources (machines, operators, IT systems and so on);
- The *Sensor Gateway* element is used to collect data from different sensors placed on the product. It is composed by one or more sensors that are able to collect product usage information. Some examples of the data gathered could be represented by temperatures, power consumption, fuel consumption, geographical location and so on.

6.4 APPLICATIONS

This section describes how LEAP interoperability standards have been implemented and used in industrial use cases related to the LinkedDesign project.

6.4.1 Production Line of Car Chassis

As described in Section 8.1, when monitoring production lines of car chassis, different actors need different information, as illustrated in Fig. 6.7: chassis parts are first moved from the oven to a press machine, and then to other operations. This process segment involves two robots to transfer the chassis

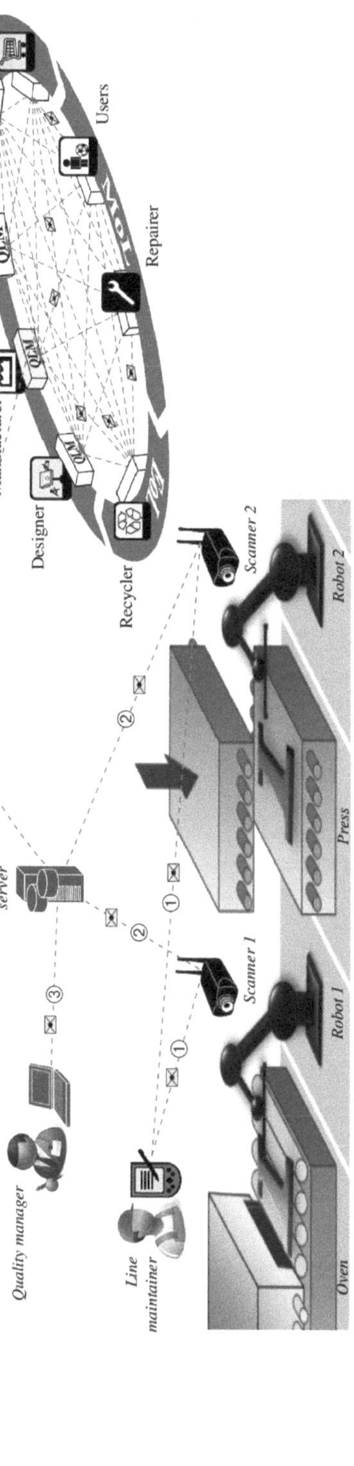

Figure 6.7 Hot stamping process implementing O-MI and O-DF for monitoring and control purposes.

part from machine to machine. On the one hand, the actors involved in the manufacturing plan expressed the need to check each chassis part throughout the hot stamping process and, on the other hand, the need to define communication strategies adapted to their own needs. Accordingly, scanners are added between each operation for the verification procedure, and O-MI/O-DF messaging is adopted to provide the types of interfaces required by each actor. O-MI nodes were installed at the relevant physical nodes of the production line, namely: the scanners; a server in charge of collecting data generated on the production line; the line maintainer's device; and the quality manager's computer.

The line maintainer expressed the need to receive all of the verification events generated by scanners 1 and 2[b] to identify in real time when a problem occurs on a chassis part. Accordingly, the maintainer directly subscribes to these scanners (see communication denoted by ① in Fig. 6.7), which is done by performing an O-MI read query by: (1) setting the interval parameter to "−1" to indicate that the subscription is *event-based*, (2) by including his own address as callback, and (3) by setting the TTL parameter to "−1" to indicate that the validity period of the subscription is "forever." These parameters are shown in lines 1, 2, and 3 of the subscription request given in Fig. 6.8. Line 8 indicates the name of the InfoItem to be subscribed to, that is, *StatusD*.

```
   <?xml version="1.0" encoding="UTF-8"?>
1  <omi:omiEnvelope xmlns:omi="omi.xsd" version="1.0" ttl="-1">
2      <omi:read msgformat= "odf.xsd" interval= "-1"
3      callback="http://207.46.130.1/omi/">
4          <omi:msg xmlns="odf.xsd" xsi:schemaLocation="odf.xsd odf.xsd">
5              <Objects>
6                  <Object type="Hotstampingmachine">
7                      <id>HotStamp1223</id>
8                      <InfoItem name= "StatusD"/>
9                  </Object>
10             </Objects>
11         </omi:msg>
12     </omi:read>
13 </omi:omiEnvelope>
```

Figure 6.8 ***Line maintainer subscribes (forever) to the InfoItem StatusD to Scanner 1.***

[b]In reality, there are two different scanners after the press.

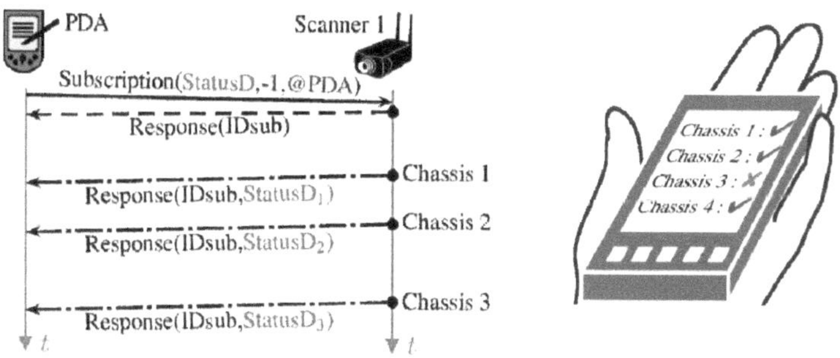

(a) (b)

Figure 6.9 ***Subscription with callback address performed by line maintainer.*** (a) Sequence diagram related to the subscription. Line maintainer's device is here Illustrated by a personal digital assistant *(PDA)*. (b) Triggering of alarm.

The sequence diagram in Fig. 6.9a gives insight into the transactions resulting from this subscription. First, a response that contains the ID of the subscription is returned to the line maintainer. Then, each time a chassis part passes under scanner 1, the subscribed InfoItem value (denoted by $StatusD_x$ in Fig. 6.9a, x being the part reference) is pushed to the line maintainer's device through a new O-MI response. Based on these events, it is possible to develop scripts, for instance, to raise an alarm if a failure occurs on a part. An example is illustrated in Fig. 6.9b where chassis part 3 has a fault.

The company server subscribes to scanner events by specifying a callback address, in a similar manner as the line maintainer (see communication denoted by ② in Fig. 6.7). Accordingly, people internal or external to the organization can access the subscribed data on the server depending on the security settings.

Unlike the line maintainer, the quality manager is not interested in receiving a continual flow of events from the scanners. His primary function is not to guarantee real time control but rather to deal with weekly or monthly evaluations (eg, to estimate the failure rate over a period of time). Accordingly, the second type of subscription supported by O-MI is more appropriate, which consists in retrieving (ie, polling) one or several historical values on the server by issuing a new O-MI read query containing the subscription ID. This corresponds to the communication denoted by ③ in Fig. 6.7. For this to happen, the quality manager sends a subscription request (via his computer), as illustrated in Fig. 6.10, by setting the interval parameter to "−1" but, this time, without including a callback address (see the argument "null" in Fig. 6.10). Then, the server is aware that the events generated by Scanner 1 should be kept locally as long as the subscription is

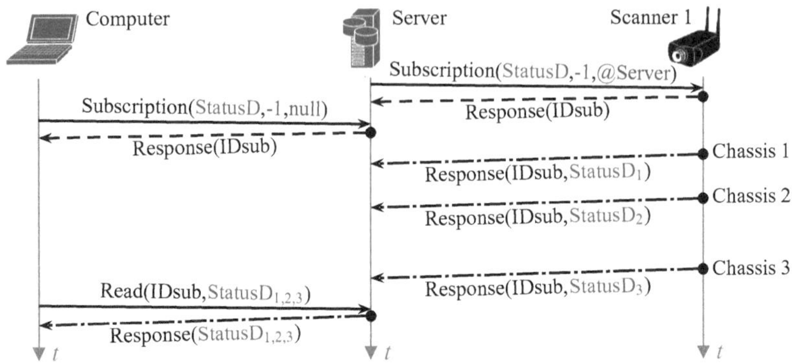

Figure 6.10 ***Subscription without callback address performed by the manager.***

valid. The quality manager can thereafter issue new read queries with the request ID of the subscription for retrieving the collected events as depicted in Fig. 6.10 (see the Read request with the arguments "IDsub" and the range of stored values denoted by $StatusD_{1,2,3}$ to respectively obtain values of chassis parts 1, 2, and 3).

If other organizations in the product lifecycle support O-MI messaging, and if security rules allow them to access particular information from the production line, then they could further use O-MI to subscribe to it, modify it, and so on. This corresponds to the communication denoted by ④ in Fig. 6.7.

6.4.2 COMAU Use Case—Collection of Data From the Field

One of the COMAU needs faced during the LinkedDesign project was the collection of data from the field. This use case is described in more detail in Section 8.2. This is a big issue for COMAU because it enables a series of competitive advantages, such as: (1) the real time monitoring of the system during the use phase, keeping under control the costs and environmental impacts generated along the use phase, besides the technical performance; (2) the possibility of updating the company database, improving the LCC and life cycle environmental impact estimations of the next proposals; and (3) the extraction of valuable knowledge from the field for the designers, enabling the comparison between real data and estimated data (during the design phase) and the understanding of the behavior (failure rate, availability, reliability, energy consumption, etc.) of the system during the use phase, in order to improve the design of next systems. The advantages are distributed in the short term, keeping both the existing system and, therefore, the customer satisfaction under control, but also in the long term improving the design of new systems.

Industrial systems are able to collect considerable amounts of data thanks to the different sensors installed on each component, functional group, and station of the line. The issue is how to export these data from the line to the decision makers, or the employees who have to manage these data. The O-MI standard can be helpful to solve this problem, enabling the extraction and exportation of data from the system's sensors to a tool (PLM Data Service), which is able to return valuable knowledge to the decision makers. Fig. 6.11 shows where O-MI can be used within the system lifecycle of COMAU. Furthermore, it highlights how data collected from the field are used on the existing system, comparing the real performance with the ones reported on the proposal, and on the next systems, improving the design capability of COMAU designers. O-MI standard enables the extraction and exportation of data from the field to the PLM Data Service. The information extracted from the operating industrial system can be different and can cover different fields, from the machine status (operating, failure, starved, blocked, etc.) to the energy consumption and the information to calculate the Overall Equipment Effectiveness (OEE). Furthermore, it is possible to choose the data granularity, choosing to have aggregated data of the machine, station, or data from the single functional group of each station.

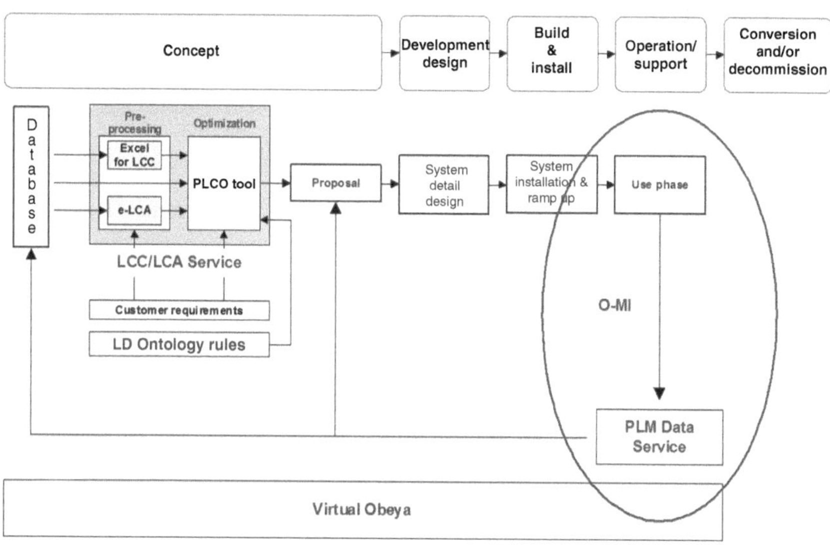

Figure 6.11 *Application of O-MI within COMAU Use Case.*

6.5 CONCLUSIONS

The aim of this chapter was to describe LEAP interoperability standards. In the first paragraph, the background behind the standards was presented, highlighting the aim of the LEAP interoperability standards to handle all the relevant information of the CL2M so that information from any single lifecycle phase may affect processes and decision making in the other phases. Section 6.2 presented the two standards developed within LinkedDesign project and The Open Group by the IoT Open Group: the O-DF (Open Data Format) standard and the O-MI (Open Messaging Interface) standard. Section 6.3 describes the functionalities and the technical developments of the related lifecycle management standards initiatives, mainly O-LM. Section 6.4 presented some of the applications of the LEAP interoperability standards. The two applications come from the industrial sector: the first one monitors the production line of a car chassis, collecting data from the field in order to have good parts; the second collects data from the field in order to keep costs and environmental impacts generated by the existing system under control, it will also help to improve the design of the next system.

The development of a lifecycle management standard is a complex process. Although O-LM and the hierarchical structure have been developed, the whole schema is still under evaluation in order to verify the pertinence in different scenarios using various products. The specification document of the O-LM is planned to be ready in 2016.

REFERENCES

[1] Kiritsis D, Bufardi A, Xirouchakis P. Research issues on product lifecycle management and information tracking using smart embedded systems. Adv Eng Inform 2003;17:189–202.

[2] The Open Group. An introduction to Quantum Lifecycle Management (QLM). White paper of QLM Work Group; 2012. 13 p.

[3] Främling K, Holmström J, Ala-Risku T, Kärkkäinen M. Product agents for handling information about physical objects. Report of Laboratory of Information Processing Science series B, TKO-B 153/03, Helsinki University of Technology; 2003.

[4] Kärkkäinen M, Holmström J, Främling K, Artto K. Intelligent products—a step towards a more effective project delivery chain. Comput Indus 2003;50:141–51.

[5] Meyer G, Främling K, Holmström J. Intelligent products: a survey. Comput Indus 2009;60:137–48.

[6] Främling K, Holmström J, Loukkola J, Nyman J, Kaustell A. Sustainable PLM through Intelligent Products. Eng Appl Artif Intell 2012;26:789–99.

[7] The Open Group. Open Messaging Interface (O-MI), an Open Group Internet of Things (IoT) Standard; 2014. Available at www.opengroup.org/bookstore/catalog/c14b.htm

[8] The Open Group. Open Data Format (O-DF), an Open Group Internet of Things (IoT) Standard; 2014. Available at www.opengroup.org/bookstore/catalog/c14a.htm
[9] Stark J. Product lifecycle management: 21st century paradigm for product realization. Berlin, Heidelberg, New York: Springer; 2011.
[10] Thoben K-D, Eschenbächer J, Jagdev H. Extended products: evolving traditional product concepts. In: seventh international conference on concurrent enterprising engineering the knowledge economy through co-operation. Bremen, Germany; June 2001. p. 429–439.
[11] Främling K, Kubler S, Buda A. Universal messaging standards for the IoT from a lifecycle management perspective. IEEE Internet Things 2014;1:319–27.
[12] Främling K, Harrison M, Brusey J, Petrow J. Requirements on unique identifiers for managing product lifecycle information—comparison of alternative approaches. Int J Comput Integr Manuf 2007;20:715–26.
[13] Främling K. Tracking of material flow by an Internet-based product data management system. Tieke (Finnish Information Society Development Centre) EDISTY magazine 2002;1:24–5. [in Finnish].
[14] Främling K, Ala-Risku T, Kärkkäinen M, Holmström J. Design patterns for managing product life cycle information. Commun ACM 2007;50:75–9.
[15] Gamma E, Helm R, Johnson R, Vlissides J. Design patterns: elements of reusable object-oriented software. Reading: Addison Wesley Publishing Company; 1995.

CHAPTER 7

LEAP Virtual Obeya

M. Rossi*, M. Cocco*, S. Terzi*, K. Kristensen**, S. Parrotta†, J. Krogstie** and D. Ahlers**

*Polytechnic of Milan, Milan, Italy
**NTNU, Trondheim, Norway
†HOLONIX S.R.L., Meda, Italy

Contents

7.1 INTRODUCTION

Today's engineering operations are largely multidisciplinary. Consequently, the discussions, clarifications, negotiations, and decisions that occur in the interfaces between different phases and disciplines in an interdisciplinary value creation process constitute a large portion of typical engineering projects. The quality of the decisions that are made largely depends on access to a decision-making basis, within which all of the relevant stakeholders have been given the opportunity to discuss potential issues and problems. The

Taking the LEAP
http://dx.doi.org/10.1016/B978-0-12-805263-1.00007-4

time in which the different actors have access to certain types of information is also of great importance, especially within manufacturing processes. Decisions made in real time are also more valuable, and these can only be effectively supported if the tools enabling access and the required information flows are available for the different involved actors.

Furthermore, the formalized knowledge that a company possesses is only to a limited extent correlated with profitability, which seems to be driven to a greater extent by the systems, work practices, and the culture for putting the knowledge to work in a collaborative manner than the characteristics of the formalized knowledge base itself.

7.2 TOWARD A GLOBAL COLLABORATIVE ENVIRONMENT

In addition to becoming more multidisciplinary, engineering is becoming more collaborative as a result of global value chains, increased specialization, more complex product–service systems, and distributed development teams. Moreover, networked knowledge exchange between engineers, both inside and outside the boundaries of projects, is becoming increasingly important. This is particularly true in improving manufacturing processes: a closer integration of the different phases in a product lifecycle assumes effective collaboration between different roles across multiple phases (designers, engineers, manufacturers, etc.), but current collaboration platforms and approaches are still fragmented and disconnected, and they do not support true lean engineering to the desired extent. As a result, they fail to effectively fill the knowledge gaps observable in current engineering systems, handling mostly structured data and information. In this chapter we will present how a new IT-based tool called LEAP Virtual Obeya tries to overcome the limitations of existing tools, extending a state of art concept and incorporating lean thinking for teams engaged in specific processes, working on specific tasks.

7.2.1 Obeya: A "Large Room"

Ōbeya (大部屋) is the Japanese word for "large room" [1,2]. It originated in the G21 project at Toyota in the 1990s, which led to the first generation Prius [1]. At the onset of this project, Takeshi Uchiyamada was appointed Chief Engineer for the project. He felt that he lacked the necessary authority to make the optimal decisions, and thought that he could be overrun by experienced discipline leaders in a way that was not optimal for the project as a whole. Therefore, he needed the support of the other discipline leaders

Figure 7.1 ***A traditional Obeya.***

whenever he had a decisive discussion with one of them [3]. To achieve this, he instituted the "large room" as an arena for all of his discussions with the discipline leaders. The other discipline leaders would be present in this room, and the documents and data would be available to all (Fig. 7.1). As the project progressed, the results were so promising that the idea was taken up by the Toyota organization and after a while was made a standard part of the Toyota Product Development System [4].

The Obeya is claimed to be beneficial in decreasing the lead-time of the development process [1]. This room is big enough for the team and it has no internal partitions or walls [5]. The most noticeable characteristic of an Obeya is that the room's walls are covered by visual documents in an A3 format, with the goals of information gathering (from the various functional groups involved in the project) and information management (due to the use of graphs, schedules with checkpoints, etc.) as shown in Fig. 7.2.

Currently, Obeyas have gained popularity as an effective "low threshold" solution with high usability for multidisciplinary engineering teams, in line with lean product development thinking. Despite this fact and the advantages outlined above, current physical Obeyas are far from being perfect. In particular, the latest IT-enabled trends in management engineering

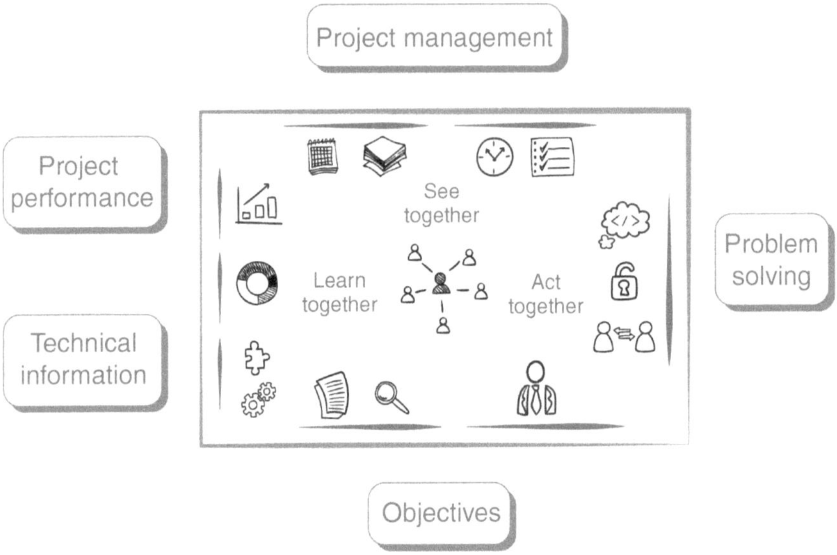

Figure 7.2 ***Typical characteristics of the traditional Obeya.***

call for the definition of new tools that, while maintaining the same ease of use and the same benefits of "classical" Obeyas, are capable of allowing a better integration of data and a tighter collaboration among people. It is now apparent that classical Obeyas, as they are typically implemented and used nowadays, lack embedded support for distributed collaboration. Due to their paper-based characteristics, they typically have no connection with the underlying information architecture of an enterprise. As a result, knowledge can be aggregated or updated only manually, it is not possible to define import/export functions, and it is not possible to implement a definite data persistence mechanism. Moreover, the gathering and the processing of information are usually handled in a fragmented, incomprehensive way. Most important of all, meetings are supposed to take place in a single physical room [6].

While many studies point to the unique benefits of being colocated, always being colocated is not a viable option for many engineering organizations today. Some important reasons for this is that organizations are geographically distributed, and teams and projects often follow the same pattern to improve competitiveness by utilizing engineering competencies in countries and regions with a lower cost base than Europe and North America. In addition, getting teams together on a regular basis through business travel is not a good option, with the exception of high-value, high-impact meetings where the benefits are higher than the combined cost of business travel and lost productive time. In addition, engineering specialists and managers have very busy agendas, and scheduling large colocated meetings is very challenging given limited availability and because the required planning time for global business travel far exceeds the desired time frame for decision making or problem solving when problems occur, leaving distributed meetings the only viable option for many situations.

A study of the limitations in existing Obeyas performed within the LinkedDesign project indicates that they typically share a set of limitations, as shown in Table 7.1 [7].

7.2.2 LEAP Virtual Obeya: A "Timeless Spaceless Room"

The LEAP Virtual Obeya is a new concept that significantly extends and enhances the functionalities of the traditional and physical Obeya. The purpose is to overcome the existing time and space limitation defining a new virtual IT-based or digital room. A main objective of the LEAP Virtual Obeya is to accelerate value creation by offering a rich, relevant, and context-specific front end supporting user collaboration across multiple tasks in a product lifecycle. The focus is on value creation, and as the portion of collaborative activities in knowledge-intensive engineering organizations is steadily increasing and waste reduction is a key theme in lean methodologies, effectively targeting and reducing waste in collaboration is regarded as an interesting, precise, and forward-looking approach to achieve a substantial reduction of overall waste in lean engineering. This is achieved by providing high usability and in-context, easy, and rapid access to aggregated knowledge, where relevance, purposefulness, and transparency remain guiding principles. To assist and guide subsequent UI research and technology development activities, requirements have been elicited based on identified limitations of existing Obeyas (see Table 7.1).

The LEAP Virtual Obeya has been designed as a fully digital, flexible management engineering front end, which rests on highly configurable information panes and in which information retrieval is semiautomatic and dependent on both context and user roles [8]. The application enables colocated, distributed or blended teams (where some team members are colocated while others are not) to work together thanks to a large-scale visualization of contents, although desktop personalized views are allowed and even useful. Mobile devices are supported but not essential to the user's experience. The tool provides access to all of the relevant sources of information, regardless of their position and context, and access to data is

Table 7.1 Requirements based on common limitations of current Obeyas

Observed limitation	Requirement(s)
Lack of distributed collaboration support.	Offer support for distributed teams working toward common repositories, avoiding duplication. Offer appropriate interaction mechanisms for distributed teams, including symmetrical or asymmetrical views, depending on role and context.
Manual knowledge aggregation only. Static, cluttered, no filtering.	Offer automated or semiautomated context- and role-based knowledge aggregation based on semantics/dynamic filtering based on relevance criteria. Interactively adapt and adjust the composition of this aggregation based on users' input/actions.
No in-context adaptive GUI provision.	Offer automated or semiautomated activity-centric GUI configuration in-context. Interactively adapt and adjust the composition of the GUI based on users' in-context input/actions.
Only passive; no or limited opportunities to interact directly with underlying information architecture.	Explore selected, high-value direct-access linkages to and interaction with underlying (reference) information architectures.
No information persistence between task boards/knowledge briefs.	As they are not dynamically updated and linked with underlying information architectures, updates done in one task board do not link to other task boards with related information. It is therefore a requirement to explore such linkages, to improve the workflow between different in-context views and task boards.
Fragmented records; development typically documented across multiple sources and formats.	A single project may have several task boards that all need to be kept up to date to cover all relevant aspects. It is therefore a requirement to explore the possibilities of unifying this information so that it can be subsequently accessed from different contexts and also across different locations in split location engineering settings.
Inefficient separation of computer-based "work" formats and physical "decision" formats. No simple import/export functions between physical (often paper-based) and virtual media formats.	To further enhance usability in real engineering settings, LinkedDesign will explore linkages between physical (often paper-based) and virtual media formats, to "soften" or eliminate separation of "work" and "decision" formats. This includes, but is not limited to an exploration of "export to paper" functionalities where the contents of a virtual task board can be easily printed to a large paper format knowledge brief (such as A3), and import or reference lookup functions where each printed A3 knowledge brief, for example, has a QR code that takes you directly to the virtual version of the same task board (for easy checking/alignment/processing).

not only passive (ie, only views supported) but also active (ie, data can be modified thanks to seamless connection to the underlying repositories) [8]. Communication can be carried out through the LEAP Virtual Obeya in a synchronous or asynchronous way: namely, a LEAP Virtual Obeya can be used to support distributed meetings or it can be shared by different people at different times in order to facilitate the overall knowledge process. The LEAP Virtual Obeya is positioned to take advantage of an extended ecosystem of related applications (widgets), which represent an enhanced version of similar functionalities offered through the "A3" sheets of classical Obeyas. Among the most important widgets to date, the most useful are those capable of performing semantic searches on predefined repositories. The other supporting applications that have been created so far have mostly centered on product lifecycle, operations, and supply chain management issues. The content files that are usually viewed during a meeting (.doc or .pdf files, .ppt, spreadsheets etc.) can be considered widgets themselves when used in an appropriate context [6].

7.2.3 Inside the Virtual Room

Realizing a virtual Obeya means realizing an alternative "room," which is not a physical room but is fully digital and can be used to enhance the selected characteristics of physical rooms designed and/or used for collaboration. The virtual Obeya must carry over most of the functionality of the physical Obeya. This means that it should allow:

- person to person communication;
- seeing each other's (digital) content;
- posting and reading posts;
- updating as new things happen;
- discussing multidisciplinary issues; and
- voting/polling.

To realize this, strong audiovisual support is required. The visual aspect means that we must have functionality for seeing other participants in real time through video/telepresence. This functionality is available in business-grade videoconferencing and telepresence systems, UC&C platforms, as well as in consumer-grade solutions like Skype or Google Hangout. Following the introduction of the browser-based real time collaboration (RTC) protocol, a number of platforms now offer full multiparty videoconferencing functionality using just a modern browser.

Depending on needs, videoconferencing can be set up from each of the participant's computer/device, or an external camera can be used to capture the full room. The former is usually better in situations where reading facial expressions adds value to the discussions, while the latter is usually better when there are a lot of movements in the room, or someone is interacting with paper-based media or working on a physical whiteboard.

The other functionality that is required is the possibility to see posted information at a later stage. This includes all kinds of information. In a traditional Obeya, information should be posted in A3 sheets, so a first step could be to realize functionality for storing, retrieving, and displaying such sheets. This is, however, a point to take advantage of the added possibilities afforded by the digital medium, so multimedia content should be handled in a simple and user friendly manner. The audio system must let participants communicate with good quality audio. This is usually not a functionality problem because this is a function of all conference systems, but it can be a quality problem as audio remains a main information carrier in knowledge-intensive discussions, and the audio quality offered by computer speakers and microphones often do not provide the quality required for a smooth experience where all of the participants are able to easily understand everything that is being said.

It must also allow spoken information to be stored (or posted) and it must allow retrieval of the information. Together, these functionality requirements specify much more than can be realized based on single persons with their own PC sitting in separate offices. They point toward the set up of a physical, "traditional" Obeya at the main site, possibly additional physical Obeyas at major remote sites, and the possibility for single persons to take part via their own PCs, possibly with some extra equipment. The physical Obeyas would not be traditional because all of the information posting should be net-based, with no physical A3s. A solution with large, full-wall displays combined with smaller displays for special purposes can be envisioned.

7.2.4 IT Architecture

Thanks to the LEAP Virtual Obeya, a number of people involved in collaborating groups can work together, both individually and as a team. The Obeya's walls are able to provide different kind of views on the meeting's topics, according to the context and the participants' role. General views are particularly appropriate for being shown on large screens, to allow everyone to contribute to the discussion, but individual views are also supported. Since the different browser sessions are all connected to the same wall instance, the system provides consistent, automatic updates of all visualizations, to keep people up to date with the most recent information. The three-layer

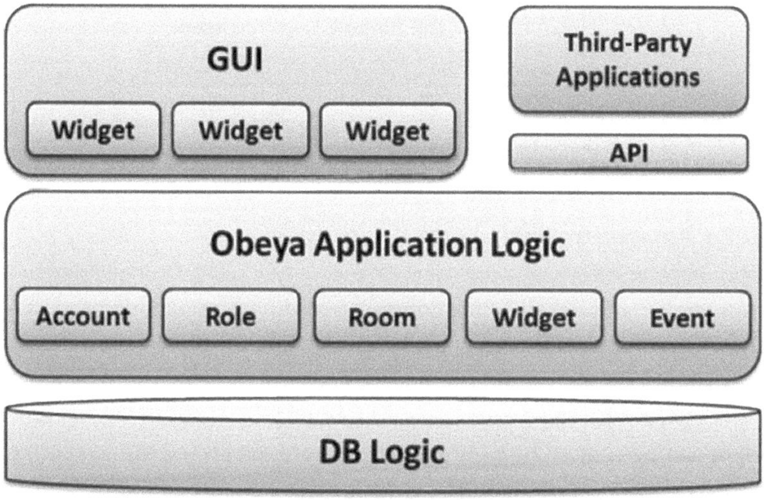

Figure 7.3 ***LEAP Virtual Obeya architecture.***

architecture shown in Fig. 7.3 that has been used for the application is the classical MVC from the computer engineering literature. A GUI superimposes on a logic layer, whose aim is to filter and manage content. Underneath the system's logic there is a set of classes which handles the state of the various objects that compose the Obeya (Room, Role, Account, etc.). The Obeya data is stored in a database, this allows every team and/or every participant to recall the meeting's contents as they like, and enables them to work on these contents synchronously or asynchronously, while in a shared meeting, in a local meeting or even individually, with the greatest freedom of choice. The LEAP Virtual Obeya will be complemented by a number of widgets or third party web applications, which can be integrated into the tool's GUI though URL or API, providing contents and functionalities (Fig. 7.3).

7.2.5 Widgets

A widget is a small application that can be installed and executed within the Obeya. A widget has the role of a transient or auxiliary application, meaning that it just occupies a portion of the Obeya wall and provides useful information fetched from tools, documents, and other websites and is displayed in place (Fig. 7.4). A widget can be embedded using an URL or exploiting the APIs. An Obeya allows us to visualize from one and up to three columns of widgets at any time, depending on the context and user requirements and permissions. One of the most important features is that

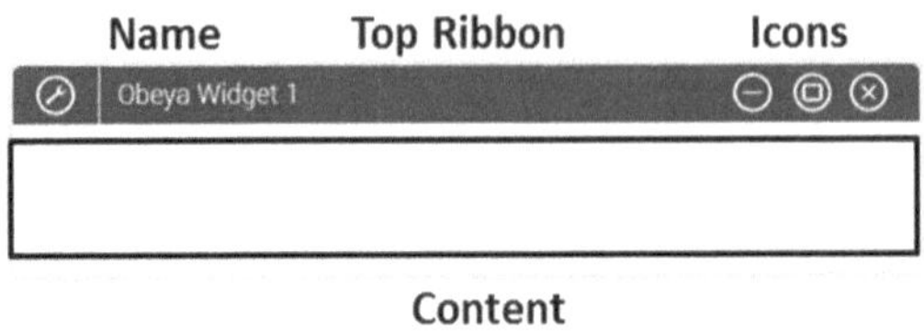

Figure 7.4 ***A widget's features.***

widgets evolve and develop throughout the whole knowledge process triggered by the Obeya, thus the widgets on the screen can be added, removed, and updated as the discussion progresses. These changes are tracked and kept in memory through the succession of different Obeya instances. In addition, a widget can be created by a template (for more details see the next section). The defined widgets come in the form of lists and, therefore, the users can generate and pick up elements in a simple and quick way for setting up new Obeyas. Although this may seem trivial, it in fact gives the tool an ease of use that other, more commercial mash-up applications do not have. Fig. 7.4 shows an example of widget layout.

7.2.6 Templates

The template is a file or a document with a standardized format used as starting point for particular application. The LEAP Virtual Obeya uses the template concept to simplify the creation of new collaborative environments.

Within the LEAP Virtual Obeya there are two types of templates:

- Widget Template, which used to replicate predefined widgets through a standardized file, document, or web-based solution; and
- Obeya Template, which is used to replicate the content of a particular and already existing collaborative environment.

A widget template is a preconfigured application that satisfies specific requirements. Widget templates with similar features can be grouped in categories. Based on an analysis of traditional Obeyas, the most noticeable characteristics have been defined as described previously in Section 7.2.1. The most used instruments have been defined for each of them, as shown in Table 7.2.

These tools are very simple and they are often composed of predefined tables and forms. Examples of these applications include A3 sheets to capture knowledge, post it walls to highlight issues, and empty white boards for open discussions represent. Typically, they are used in real rooms facilitating

Table 7.2 Most used tools within traditional Obeya

Category	Tools
Objectives	Project Objectives
	Customer Value
	Voice of customer
Technical Information	Target Products
	Drawings
	Specifications
Project Performance	KPIs (Cost, Quality, ...)
	Measurements
Project Management	Macro plan
	Work plan matrix
	Task work
	Team plans
Problem Solving	A3
	Issue Board
	PDCA

the interaction between users. Web-based solutions can be used within the LEAP Virtual Obeya to replicate tools used in physical environments going beyond the traditional Obeya limits. This can move the physical collaboration within a virtual environment supporting, for instance, split location meetings and asynchronous interaction. Most of the tools can be easily implemented using Microsoft Office 365 or Google Apps for work. Both solutions allow us to create spreadsheets, empty notes, slides, and so on. Using these platforms, the most used tools have been tested. For instance, for problem solving the most used tool is the so-called issue board. In a physical Obeya, the issue board is a table that contains some information that is edited manually by the users. This table could easily be replicated in an enhanced digital format using, for instance, Google sheets.

At least one widget template has been developed for each tool that was previously identified and others can be easily created by the users.

The Obeya Template represents the possibility to save the status of a particular room created within the LEAP Virtual Obeya. The saved state includes the widgets' contents, parameters, and position within the application. Each Obeya Template gives us the possibility to replicate the same saved Obeya for new collaborative sessions, significantly decreasing the time needed to setup a new virtual collaborative environment. This feature is very useful, especially for recurring meetings where the same applications are used several times.

7.2.7 Positioning of the LEAP Virtual Obeya

In line with lean principles, one of the objective's criteria has been LEAP Virtual Obeya's ability to aggregate and provide knowledge and information that can be characterized as the essence of engineering: "What is most important for us as an engineering team, in this situation, given these circumstances, right now?" Through exploring relevant, purposeful, and transparent UI concepts, the LEAP Virtual Obeya front end has been designed as an instrument to reduce mental overload and to free up time for value-added thinking. Knowledge workers often have to spend mental capacity (that could otherwise be invested in value-adding activities) on configuring their own workspace to make it suitable for what they are doing, resulting in lost time on task.

Each "shift" or "transition" between activities with different objectives, process inputs, and resources forces the worker to focus on the technology and process setup rather than the objectives at hand. Since relevant information in one context or work stream may be very different from what is needed in the next context, and the number of such shifts or transitions can be substantial in any single day, the use of simple configuration mechanisms that could simplify the user interface not only in a single context but across different contexts and work processes could considerably improve team robustness as well as individual knowledge worker effectiveness and efficiency.

Based on these objectives and principles, we will briefly discuss how the concept of the LEAP Virtual Obeya fits into a larger picture of related technology and tools, and we will also describe its distinguishing features. We refer to [7] for a discussion of related work regarding collaboration theories, especially collaboration in distributed/split location settings.

Growing from the idea of the Obeya as a physical room, the LEAP Virtual Obeya has been designed to be a more dynamic and productive environment supporting a wider variety of contexts and work situations. For example, documents shown in the LEAP Virtual Obeya can directly reflect changes to show if they have been updated and do not need a user interaction. In addition, given the proper frames, such as Google Docs, changes can be made directly in the view inside the LEAP Virtual Obeya. However, the LEAP Virtual Obeya is envisioned as an environment for information display, presenting and organizing information snippets or sources. It offers cascading levels of interactivity—both in terms of the configurability and the interaction logic of the LEAP Virtual Obeya itself, and on a widget level. As such, it is not only capable of showing static information but is also suitable for dynamic information sources or web-based applications. These dynamic information sources can be naturally integrated and can also be run on top of or integrated into other systems.

The main issue that LEAP Virtual Obeya is trying to solve is that although documents are available on the web and the browser is becoming the new desktop, it remains difficult to arrange information sources in a goal-oriented layout as each page is displayed 'full-screen' inside the browser. This difficulty of reproducibly handle side-by-side document layouts or even higher number of documents can be solved in a simple and straightforward way with the LEAP Virtual Obeya. The LEAP Virtual Obeya is a possible standalone tool integrating arbitrary sources, as long as they are reachable as a Web page and fulfill the minimum requirements. For split location collaboration, it gives real time updates. A change in the layout or content of the frames will be reflected in all of the other VOs opened by other users. It is thus not only a single wall but also any number of distributed walls. When we look at other solutions, limited support exists for placing the windows of desktop applications. However, there are tools which allow us to arrange the windows in a certain layout, or such functionality is built into the OS. Yet this only works on currently open applications for the current user. These tools usually cannot open applications and windows, and they are not designed to exchange layout data. Similarly, office tools and others support side-by-side document comparison. We also see web Desktops trying to replicate the user interfaces of desktop operating systems, which serve a different purpose but may share a similar technology basis.

Toward a workspace concept, integrated development environments (IDEs) are highly integrated programming tools for a specific task, have access to all relevant project documents, and allow the definition of layouts for different software development tasks within the workflow [9]. We can think of the LEAP Virtual Obeya as taking some ideas from this and developing an IDE for document and application presentation and collaboration in an Obeya-type professional work environment.

The engineering-oriented LEAP Virtual Obeya has a counterpart in the management world. Business dashboards show key business performance indicators across the enterprise in one interface. They focus on data access from different data sources in the enterprise as a decision support and management information tool [10]. They put an emphasis on design for the communication of condensed overviews [11].

We have an example of a dashboard in the automotive use case in Chapter 8.2 (cf Fig. 7.7), which is shown inside a LEAP Virtual Obeya. Similar in name, Apple Dashboard in MacOS serves as a secondary desktop, and many similar tools or widget engines for different operating systems have been proposed. Apple Dashboard deals with the arrangement of widgets, which are self-contained lightweight single-purpose apps or embedded snippets of web pages. In this case, the focus is on providing small snippets of specific information on a local computer. All of these widgets are apps that need to be specially programmed. On the other hand, although the frames in the LEAP Virtual Obeya show only web content, they can integrate any type of web documents, including existing or newly developed web apps. The different desktop widget engines allow arbitrary placement of widgets, while the LEAP Virtual Obeya follows a flexible grid where widgets can be arranged in a single, double, or triple column setup depending on objectives and team preferences.

On the other side of the spectrum we find large scale command and control centers; for example, for use in network operation centers that visualize the output of multiple systems on a large wall. This is usually a hardware-based solution that shows the video output of multiple workstations or other video sources, and it is often driven by multiscreen or multiprojector setups to scale [12]. We also find large interactive multiperson multitouch walls that support same location collaboration but which usually require specialized software or at least an adapted operating system to work [13]. Although the LEAP Virtual Obeya does not support such scenarios, it comes closer to the operations center in that it benefits from a large projector wall that replicates the original scenario of the Obeya.

Finally, SharePoint is another cooperative web-based application. While they are similar, we do not view them as competitors but rather as complementing approaches. The LEAP Virtual Obeya provides the Obeya experience, in the spirit of lean management practices. It allows fast definition of new walls and changing of views. It opens the possibility of breaking down collaboration barriers within and between collaborative teams, and elevates the traditional Obeya concept into a timeless and spaceless context. One of the features of the LEAP Virtual Obeya is that it can be embedded into any web portal (SharePoint included). In itself, it can also integrate portal components and also web-based apps, including any chat or RTC tool that runs inside a browser. This is a valuable degree of freedom because it enables the application to leverage a portal's peculiar features (such as document management) and, at the same time, it makes it independent from any specific technology platform.

Finally, the LEAP Virtual Obeya is a Lean Management tool. It is lightweight, in line with the lean methodology, is web-based and thus easily accessible without high technology or software barriers, provides persistence of layouts, real time synchronization between users, and information visualization and access.

7.3 THE LEAP VIRTUAL OBEYA IN THE REAL WORLD—USE CASES

Within the LinkedDesign project, the LEAP Virtual Obeya was used and evaluated in certain industrial use cases (see Chapter 8). The project used both the standard LEAP Virtual Obeya interface and specifically designed widgets adapted to the use case and integrating industrial data sources. The following gives a brief overview of these specialized installations.

7.3.1 Assembly Line Provider Case

The LEAP Virtual Obeya represents the LEAP user interface. In the Assembly Line framework, it is used to exploit the possibility of the PLM service to record event-based data; it also provides large scale data sharing capability (see Chapter 8.3). Comau Powertrain, the business unit involved, has to combine multiple efforts from different departments of the company located in different geographical areas. A collaboration tool such as the LEAP Virtual Obeya could significantly help the designer in sharing enriched information and the possibility to connect widgets to the Obeya allows full interoperability between the various design teams.

On the other hand, this tool can be extremely useful during the ramp-up phase, when the machines are installed in the customer's plant, and are subject to adjustments and set up procedures. In this phase, any issue arising in the machine might cause significant delay, so it must be addressed quickly by rapidly involving the right people. For example, if a failure occurs in a machine, the "failure" event is registered by the PLM and associated with a predetermined failure cause. This information is used to update the failure rate database for that kind of machine and to refine the associated LCC calculation. On the other hand, according to the Comau or the customer specification, the same information can be sent directly to the individual responsible for the production or to the Comau personnel in the plant through a notification in the LEAP. This allows the person responsible for the production to be warned in real time and plan an intervention to restore the production. An example of LEAP Virtual Obeya used within Comau is shown in Fig. 7.5.

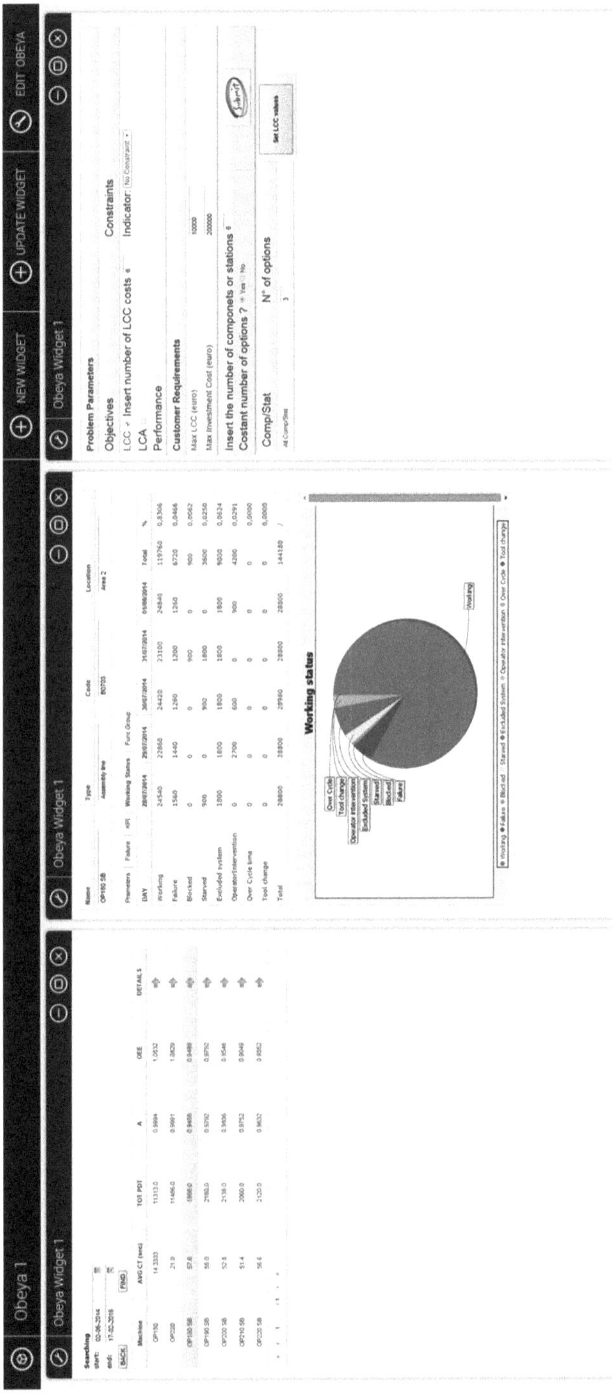

Figure 7.5 *Assembly line LEAP Virtual Obeya.*

7.3.2 Oil and Gas Case

One of the main aims of Aker Solutions (see Chapter 8.4) is to allow engineers to share information between colleagues and set up virtual environments for knowledge sharing. For a similar reason, Aker Solutions has implemented a tool named Knowledge Arena, which was created with the aim to make all information, know-how, experience, and knowledge available for everyone in the company, independent of location or time-zone. This tool was developed using SharePoint, a Microsoft collaboration and data repository tool used in Aker Solutions. Consequently, we have evaluated the possibility to also use the LEAP Virtual Obeya inside SharePoint, investigating the benefits and problems of this solution. The LEAP Virtual Obeya can be embedded in SharePoint as a tool to facilitate the addition of new content that can then be made available to colleagues. In SharePoint, an authorized user can set up and embed content on internal collaborative sites. However, preparing customized sites requires a lot of time and needs to be done by skilled users. One of the advantages of the LEAP Virtual Obeya is the ability to create collaborative rooms to easily add new web content, tools, documents, or files. Moreover, the LEAP Virtual Obeya could also be used as a tool to support internal physical meetings (Fig. 7.6). A meeting's moderator can easily set up an Obeya with all of the documents that he or she would like to discuss, including presentations, documentations, PDFs, CAD files, links to internal documentation, and so on. Moreover, it is possible to link to external tools that are typically used during meetings, for instance Project Management and Visual Management tools and charts to verify a project's progress, delays, and critical situations. Other tools can be used to take minutes at meetings, create mind maps, plan the next activities, and so on.

7.3.3 Automotive Case

Within the Volkswagen (see Chapter 8.2) use case the LEAP Virtual Obeya has been used as shop floor management tool to monitor the status of production and visualizing real time warnings (see Fig. 7.7). In particular, a Process Line Obeya has been created, which is a virtual room where end users are able to see simultaneously on one big screen all of the information provided within ad-hoc widgets developed for this use case. The Process Line Obeya contains applications to monitor the production in real time. The Real Time Monitor is a tool that is used to visualize real time data collected by sensors across the whole line that sends a warning if the temperatures,

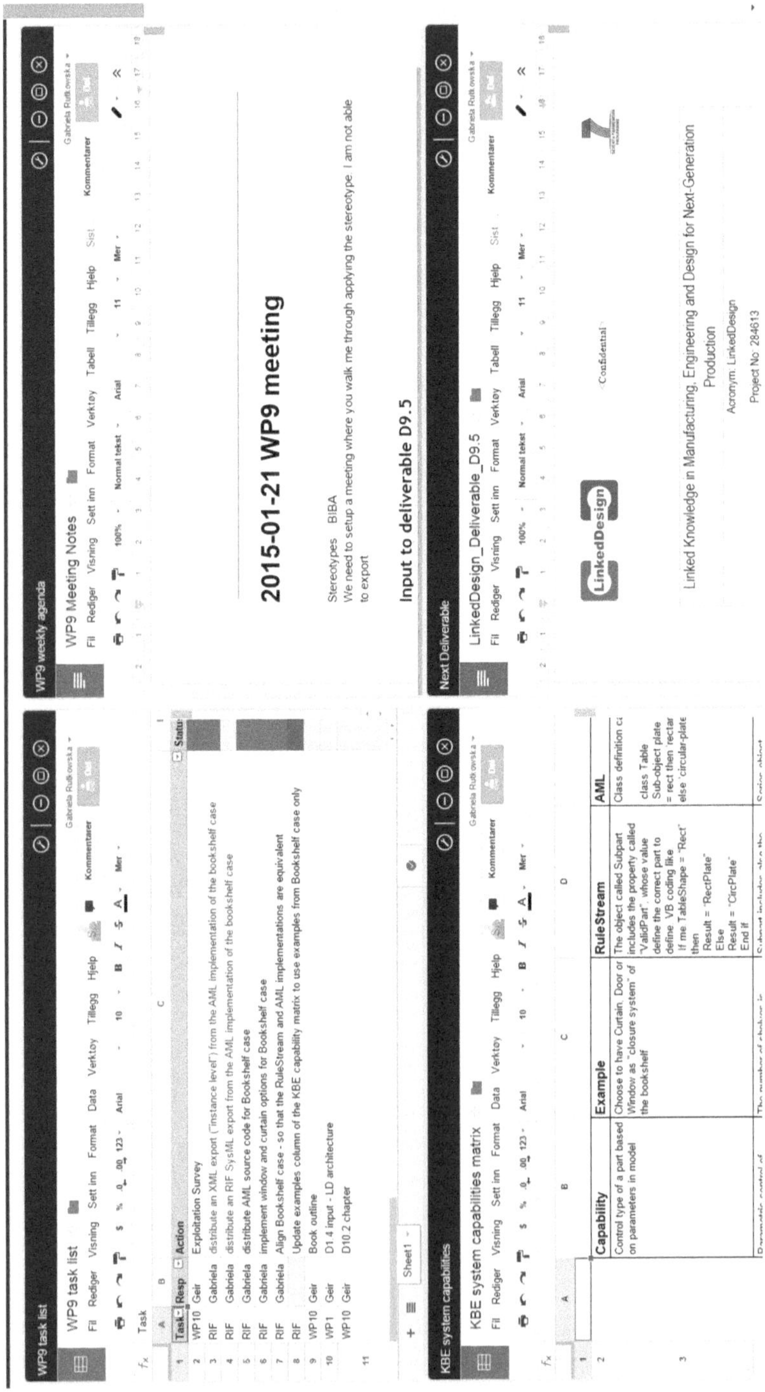

Figure 7.6 ***Oil and gas LEAP Virtual Obeya.***

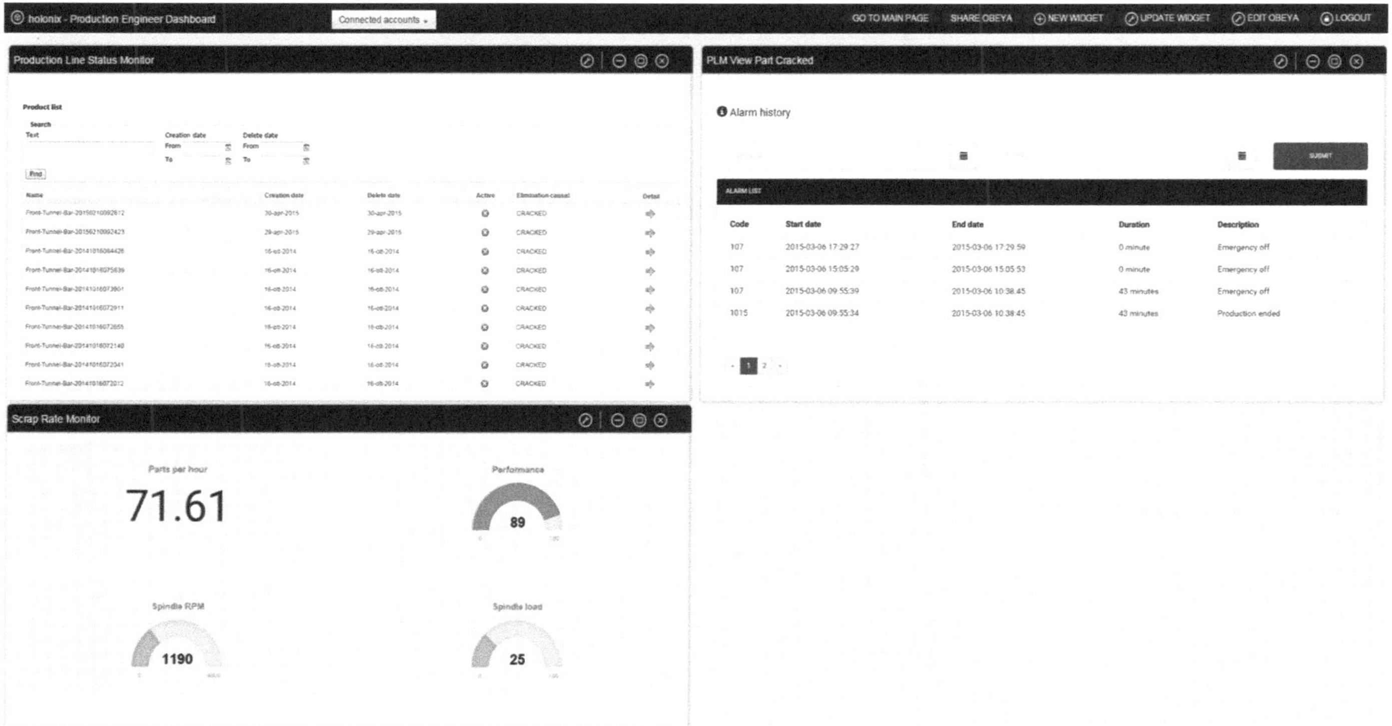

Figure 7.7 ***Automotive LEAP Virtual Obeya.*** None of the numbers are real internal data sets but are illustrative dummy data that are used for publication.

pressure, or other parameters fall out of the range. The PLM Data Service is another application that allows the system to show historical information for each manufactured product, including the sensor parameters, data of the rejected parts, and geometric deviation. Another widget allows us to show historical warnings from different machines located on the line, showing time, duration, and comments by the technical experts that fixed the failure. Finally, the last two applications show KPI on the line's performance, such as production information, scrap rate, and cycle time during the day, week, and month.

7.4 EVALUATION

The LEAP Virtual Obeya was designed to provide the interface within the LEAP architecture. As such, the evaluation of functionality and usability is mostly driven from the user side and the integration into workflows.

The employed methodology follows a selection of evaluation methods over the project lifetime. First, there are on-going evaluations and feedback to the development team, as well as an initial collection of user requirements (partly within the LEAP Virtual Obeya). Refinements are then driven by the demonstrators of the project and by the use of the LEAP Virtual Obeya within the project as a tool to support teleconferences and documentation preparation. This ensures that the project members are integrated into the development and can give their feedback.

We will next briefly report the results of two types of evaluations. The first is an evaluation within selected use cases. The second is a formal evaluation at the project finish to prepare for further development, dissemination, integration, and exploitation efforts. To do this, we will use a survey of all the project members who have used the LEAP Virtual Obeya in some capacity. The LEAP Virtual Obeya has been evaluated both qualitatively and quantitatively.

7.4.1 Use Case Evaluation of the LEAP Virtual Obeya

In LinkedDesign, the individual use cases were evaluated as a whole, their components were also evaluated. We will only briefly summarize the findings relating to the LEAP Virtual Obeya here. We chose to examine two use cases in detail. The LEAP Virtual Obeya in the project has primarily been used for two types of activities: supporting multisite meetings and collaborative document production, besides being the interface for the LEAP architecture (as shown in Section 7.4 and Chapter 1).

The LEAP Virtual Obeya demonstrators at Aker Solutions have been evaluated through test use and presentations to potential users. The LEAP Virtual Obeya was additionally evaluated through a usability expert review [14]. The LEAP Virtual Obeya has been tested in two types of meetings: first, a split location videoconference meeting between a design team and a domain expert in a different country; and second, a recurring weekly project status meeting. Finally, the LEAP Virtual Obeya was demonstrated to KBE practitioners.

The results indicate that the LEAP Virtual Obeya has the potential to improve split location engineering collaboration within the company, as well as to improve efficiency in meetings, and it has the potential to improve the quality of decision making. The possibility to collaborate on K-Briefs during a LEAP Virtual Obeya session will potentially simplify knowledge capture and experience transfer across locations. More than half of the KBE practitioners affirmed that the LEAP Virtual Obeya would be useful to them in their work on a conceptual level. The main remaining challenges for the LEAP Virtual Obeya are integrating and aligning it with existing tools in the company.

At Comau, the focus was on a questionnaire that qualitatively evaluated three main developed tools. The LEAP Virtual Obeya was investigated regarding its capability to enable collaboration between different company departments, the reduction of the time and cost spent for the coordination, the ability to search a project and the related information, and questions relating to project management and collaboration. The evaluation results in general proved to be a positive ranking for the project objective, although some concerns were raised about integration. The LEAP Virtual Obeya was considered to involve all of the functional roles in a web-based collaboration environment. This fundamental feature was assessed as both an advantage and a disadvantage. In fact, in this way, it is possible to have a broad and easy usage, but it is difficult to guarantee the confidentiality of the content.

7.4.2 Comprehensive LEAP Virtual Obeya Questionnaire

To get a broad feedback on the experience of the use of the LEAP Virtual Obeya, a questionnaire was developed to capture input from the different users of the LEAP Virtual Obeya from the extended project. The survey consisted of the following parts:

- Organizational background and the use of visual means for communication across the organization. The questions were based on the work of [15].

- Questions following the technology acceptance model (TAM), investigating the perceived usefulness (PU), perceived usability, and intention to use (IU) [16].
- Specific questions relative to the usability of generally managing and using the LEAP Virtual Obeya.
- Questions investigating the benefit of the LEAP Virtual Obeya for specific tasks, including project management, knowledge management, collaborative work, and cross-organizational communication.

We will next present the main results from this survey. We received 23 responses to the survey from people related to the project that had been using the LEAP Virtual Obeya for different tasks over varying periods of time. This includes both people working on the project and people being involved in use cases or courses, who were not primarily working on the project. All of the responses were provided in January 2015, based on the use of the version of the LEAP Virtual Obeya available at that time (not including the latest additions of functionality, including roles and templates). The scale range was 1 (Strongly disagree), 2 (disagree), 3 (neutral), 4 (agree), 5 (strongly agree) (ie, so-called Likert scales).

For an overview of the participant's organizational background and the use of the LEAP Virtual Obeya, the following questions were included:

- The use of simple and visual communication is anchored in the company culture.
- There is a culture in the organization for sharing information and knowledge.
- The use of many different design tools makes collaboration difficult.
- It is easy to find the right information to do my work in the organization.

The results show that there is clearly an information sharing culture in the organizations, whereas to what extent using visual computing and issues with collaboration difficulties and finding the right information differs more.

The next set of questions asked about technology acceptance, following the TAM.

The first topic was Perceived Usefulness (PU), with the following subtopics (as shown in Table 7.3):

- PU1: The LEAP Virtual Obeya would be useful for me in my work;
- PU2: Using the LEAP Virtual Obeya would increase the efficiency of my daily work;
- PU3 The LEAP Virtual Obeya would allow me to retrieve relevant information in my work;

Table 7.3 Descriptive Statistics of questions on perceived usefulness

	N	Minimum	Maximum	Mean	Std. Deviation
PU1	23	2	5	3.65	0.775
PU2	23	2	4	3.57	0.728
PU3	23	2	5	3.43	0.728
PU4	23	2	5	3.78	0.736
PU	23	2.50	4.50	3.6087	0.52129
Valid N (listwise)	23				

- PU4: The LEAP Virtual Obeya would allow me to better plan and organize my activities.

All range from Disagree to Strongly Agree (except on the efficiency on a daily basis). Participants opined that the LEAP Virtual Obeya would be useful in around 70% of cases.

The second topic was Perceived Ease of Use (PEOU), with following subtopics:

- PEOU1: The overall interaction with the LEAP Virtual Obeya is clear and understandable;
- PEOU2: The overall interaction with the LEAP Virtual Obeya does not require a lot of mental effort;
- PEOU3: I find the LEAP Virtual Obeya easy to use;
- PEOU4: I find it easy to get the LEAP Virtual Obeya to do what I want it to do.

The average rating on PEOU1-PEOU3 was at the same level as seen on the PU-question. On PEOU4 'getting the LEAP Virtual Obeya to do what the user wants it to do', lower marks were given.

A main aspect of the TAM is the Intention to Use (IU) (as shown in (Table 7.4):

- IU1: Assuming I have access to the LEAP Virtual Obeya, I intend to use it;
- IU2: Given that I have access to the LEAP Virtual Obeya, I predict that I would use it.

Table 7.4 Descriptive Statistics on questions on intention to use the LEAP Virtual Obeya

	N	Minimum	Maximum	Mean	Std. Deviation
IU1	23	2	5	3.74	0.864
IU2	23	2	5	3.65	0.885
IU	23	2.0	5.0	3.696	0.8358
Valid N (listwise)	23				

The IU is slightly higher on average than the PU and PEOU, which shows that participants could see value in the use of the LEAP Virtual Obeya.

To test the reliability and validity of each construct in TAM, the Internal Consistency of Reliability (ICR) of each construct was tested with Cronbach's Alpha coefficient. The Cronbach's Alpha values range from 0.654 to 0.905. According to previous research [17], a reliability coefficient of 0.6 is marked as a lowest acceptable limit for Cronbach's Alpha for exploratory research. [18] also suggested that an alpha score of 0.6 is generally acceptable. All of the Cronbach's Alpha values of the constructs in our model are above the threshold of 0.6. Consequently, the scales were deemed acceptable to continue. Looking at the relationships between these variables, we find that PEOU correlates with PU and PU correlates with IU ($p < 0.001$), following the main relationships in the TAM model.

A factor that can influence the PEOU, PU, and IU is how much the respondent has used the LEAP Virtual Obeya. Differentiating between those respondents that have used the LEAP Virtual Obeya for less than one day and those who have used it for longer, we find that PU appears to decrease with use. Primarily, this relates to the fact that those who have used it only a little, have tended to use it in meetings that have been set up by others. The one exception is for perceived efficiency (PU2).

The same pattern is observed with PEOU and IU. With as few as 23 responses, we should not expect to find significant differences, but there is a tendency that those respondents with more experience also find the LEAP Virtual Obeya less useful, less easy to use, and are less likely to continue using the LEAP Virtual Obeya. This may happen because the LEAP Virtual Obeya at the current stage supports easier collaboration tasks better than more complex use areas. Note that this survey was conducted before the roles and templates functionality was introduced, which should make it easier to setup and administer more complex working environments.

In addition to the standard TAM usability questions, some additional, specific usability questions were provided to guide future development, as follows:

- The setting up of a new Obeya from scratch is easy and does not require too much time;
- The steps necessary to create a new widget are clear and understandable;
- The main page of the LEAP Virtual Obeya is simple and the buttons are understandable;
- Within a specific Obeya, the top ribbon features are clear and understandable.

Table 7.5 Descriptive Statistics for the different usage areas

	N	Minimum	Maximum	Mean	Std. Deviation
CUSTDES	11	3	5	4.18	0.751
COMDEP	21	3	5	4.05	0.590
COMLEV	16	3	5	4.06	0.443
WORK	21	1	5	3.52	1.078
DISTCOLL	22	3	5	4.05	0.486
VOKM	20	2	4	3.15	0.813
VOCON	20	2	5	3.15	0.875
VOMEET	22	2	5	3.50	0.964
VOPM	19	2	5	3.79	0.787
Valid *N* (listwise)	9				

We got answers ranging from disagree to strong agreement, with the average toward agreement, although the need for some future development could be identified.

A number of task-oriented questions were asked about the applicability of the LEAP Virtual Obeya for different tasks. If the respondents had not used it in this way, then they could answer 'not applicable', which explains why $N < 23$ (see Table 7.5). The statements were:

- *CUSTDES*: The LEAP Virtual Obeya supports the efficient communication between customers and designers;
- *COMDEP*: The LEAP Virtual Obeya supports the efficient communication between people working in different departments;
- *COMLEV*: The LEAP Virtual Obeya supports the efficient communication between organizational levels;
- *WORK*: The use of the LEAP Virtual Obeya supports elimination of unnecessary work;
- *DOSTCOLL*: The LEAP Virtual Obeya supports distributed collaborative work well;
- *VOKM*: The LEAP Virtual Obeya supports knowledge management well;
- *VOCON*: The LEAP Virtual Obeya provides the necessary information based on the context;
- *VOMEET*: The LEAP Virtual Obeya replaces traditional meetings in an efficient manner;
- *VOPM*: The LEAP Virtual Obeya supports project management well.

Here we see some marked differences, especially in tasks that support communication and collaboration across traditional organizational boundaries, which are reported to be well supported. The support of the LEAP Virtual Obeya for areas such as knowledge management and in a context-based way was less positively perceived, at least in the current version of the Obeya.

The previous part looked at the descriptive statistics for the individual questions. We will next look at the correlations between the Likert items to examine relationships between questions beyond the standard TAM-investigation. For this, we use a pairwise comparison of all items. For each pair, we compute the correlation coefficient. We use Pearson's r as a measure of linear correlation. The results lie in the interval between -1 and 1. Additionally, we compute the significance as a p-value ≤ 0.05. We see that finding the LEAP Virtual Obeya useful correlates well with a number of other factors, significantly with ease to adapt, intention and prediction to use, and the LEAP Virtual Obeya use for context-based information and support of knowledge management. We find a similar result for the LEAP Virtual Obeya being able to increase the efficiency of the work. However, we see a negative correlation for using the LEAP Virtual Obeya to find relevant information and to better plan work. We speculate that this is due to many respondents using the LEAP Virtual Obeya mostly in a meeting setting where these factors do not play a major role. The perceived ease of adaptation correlates with the use for elimination of work, distance collaboration, providing of information, and the replacement of meetings. The LEAP Virtual Obeya interface questions show that, in most cases, there is agreement between the usability of different components, with an inverse relationship between the LEAP Virtual Obeya setup and its main screen. There is further agreement of the LEAP Virtual Obeya use for project management and respondents use cases of communication within an organization. Furthermore, the intention of use correlates with the elimination of unnecessary work, providing context-based information, and the ability to replace meetings. The mostly positive correlation shows that respondents mainly identify the LEAP Virtual Obeya as consistent.

7.5 CONCLUSIONS

A main objective of the LEAP Virtual Obeya has been to accelerate value creation by offering a rich, relevant, and context-specific front end supporting user collaboration across multiple tasks in a product lifecycle. The focus is on value creation, and because the portion of collaborative activities in knowledge-intensive businesses is steadily increasing, effectively targeting, and reducing waste in collaboration is regarded as an interesting, precise, and forward-looking approach to reducing overall waste in lean engineering. This is achieved by providing high usability and in-context, easy, and rapid access to aggregated knowledge levels. Relevance, purposefulness, and transparency have served as important guiding principles.

To this end, the LEAP Virtual Obeya can be considered as a tool that offers a significant performance improvement potential for lean managers by providing an electronic, web-based version of traditional physical lean management Obeyas. Offering enhanced functionalities for engineering teams working on specific tasks, the LEAP Virtual Obeya is a mash-up application in which a number of widgets can be integrated to replace former A3 sheets of paper. Today's projects involve different groups who are dispersed all over the world, which creates a need for a virtual replacement of the physical room. This requires a tool that can be accessed everywhere and which fosters the same collaboration performance as "in person" meetings. The LEAP Virtual Obeya demonstrates that covering this open gap is possible.

Through the LEAP Virtual Obeya, one can gain a management overview of the topics being discussed in a meeting, while at the same time it allows us to get personalized views on the different issues. This enables real time and synchronous collaboration and information sharing in distributed teams. Knowledge discovery facilities are supported, and information can be modified on the fly as it is visualized and discussed thanks to seamless connections with local and cloud repositories.

It should be noted that preparation of suitable content items and of a proper physical environment can prove critical for the application to be successfully used in real lean engineering settings. The addition of new tools, in particular about project management, will be useful for competitiveness. Further practical tests with users will help to refine and improve the system, continuing cycles of development, pursuing the final goal to get a good reaction from the market. The first tests are optimistic.

An on-going evaluation with real users is important in software development. In LinkedDesign, there was a possibility to have project participants using the developed tools, in this case the LEAP Virtual Obeya, during real work situations, such as different types of meetings and collaborative document creation. Such settings can be used to gather valuable feedback. For example, based on the feedback of the partners and from the use cases, it became clear that it is necessary to integrate the LEAP Virtual Obeya deeper into existing tools and enterprise software landscapes. On the other hand, the questionnaire closes the LinkedDesign activities, but will be the basis for further development of the tools.

In conclusion, the conceptual approach of the LEAP Virtual Obeya was mostly received positively; further work will concern its exploitation and usability. LEAP Virtual Obeya has the potential to become a useful tool supporting communication and collaboration across traditional organizational boundaries to support lean collaborative methods.

REFERENCES

[1] Morgan JM, Liker JK. The Toyota product development system. New York: Productivity Press; 2006.

[2] Tanaka T. Digital Obeya. Lean summit 2012; 2012.

[3] Aasland K, Blankenburg D. An analysis of the uses and properties of the Obeya. 18th international CE conference on eEngineering, technology and innovation (ICE); 2012, p. 1–10.

[4] Osono E, Shimizu N, Takeuchi H, Dorton JK, Bigelow SJ. Extreme Toyota: Radical contradictions that drive success at the world's best manufacturer. Hoboken, NJ: John Wiley & Sons; 2008.

[5] Larman C, Vodde B. Lean primer, 2009; 2014.

[6] Terenghi F, Cassina J, Kristensen K, Terzi S. Virtual Obeya: a new collaborative web application for running lean management workshops. 2014 international ICE conference on engineering, technology and innovation (ICE); 2014, p. 1–7.

[7] Kristensen K, Aasland KE, Blankenburg D, Sivertsen OI. Virtual Obeya: collaborative tools and approaches addressing sources of waste in lean engineering. 19th ICE & IEEE-ITMC international conference on responsible innovation and entrepreneurship; 2013.

[8] Blankenburg D, Kristensen K, Aasland KE, Sivertsen OI. Virtual Obeya: collaborative tools and approaches to boost the use of simulators in concept design. Proceedings of the 27th European conference on modelling and simulation (ECMS); 2013.

[9] Kats LC, Vogelij RG, Kalleberg KT, Visser EI. Software development environments on the web: a research agenda. Proceedings of the 11th SIGPLAN symposium on new ideas, new paradigms, and reflections on programming and software (ACM Onward); 2012, p. 99–116.

[10] Eckerson WW. Performance dashboards: measuring, monitoring, and managing your business. New York, NY: John Wiley & Sons; 2010.

[11] Few S. Information dashboard design: displaying data for at-a-glance monitoring. 2nd ed. Burlingame, CA: Analytics Press; 2006.

[12] Ni T, Schmidt GS, Staadt OG, Livingston MA, Ball R, May R. A survey of large high-resolution display technologies, techniques, and applications. Virtual reality conference. IEEE; 2006, p. 223–236.

[13] Wei B, Silva C, Koutsofios E, Krishnan S, North S. Visualization research with large displays. IEEE Comput. Graph. Appl 2000;20(4):50–4.

[14] Hollingsed T, Novick DG. Usability inspection methods after 15 years of research and practice. Proceedings of the 25th annual ACM international conference on design of communication; 2007, p. 249–255.

[15] Welo T. On the application of lean principles in Product De-velopment: a commentary on models and practices. Int J Prod Dev 2011;13(4):316–43.

[16] Davis FD. Perceived usefulness, perceived ease of use and user ac-ceptance of information technology. MIS Q 1989;13:319–40.
[17] Robinson JP, Shaver PR, Wrightsman LS. Criteria for scale selections and evaluation. San Diego, CA: Academic Press; 1991.
[18] Moss S, Prosser H, Costello H, Al E. Reliability and validity of the PAS-ADD Checklist for detecting psychiatric disorders in adults with intellectual disability. J Intellect Disabil Res 1998;42(2):173–83.

CHAPTER 8

LEAP Use Cases

A. Milicic*, S. El Kadiri*, F. Perales**, S. Parrotta† and G. Iversen‡
*EPFL, Lausanne, Switzerland
**Trimek S.A., Spain
†HOLONIX S.R.L., Meda, Italy
‡Aker Solutions AS, Norway

Contents

8.1 INTRODUCTION

All of the concepts and components developed within the LinkedDesign project will be continuously evaluated using the following three representative application scenarios:

- Metrology-driven Manufacturing in the Automotive Industry—Volkswagen in collaboration with TRIMEK
- Manufacturing Design based on Plant Lifecycle Costs—Comau
- Knowledge Reuse and Collaboration in Automated Design for Offshore Engineering—Aker Solutions

While the first scenario addresses the design of the manufacturing processes, the second scenario aims to optimize the design of products. The third scenario aims to increase the efficiency of engineering projects through systematic capture and reuse of design knowledge and design automation,

Taking the LEAP
http://dx.doi.org/10.1016/B978-0-12-805263-1.00008-6

as well as the increased performance of collaborative processes. By an early prototyping and iterative improvement of the LinkedDesign solution, the project developments were aligned with real-world business needs, especially from engineering and manufacturing. The three application scenarios are described in detail in following sections.

8.2 METROLOGY-DRIVEN MANUFACTURING IN THE AUTOMOTIVE INDUSTRY

The need for a customizable user interface with respect to the demands of different user groups were defined in the beginning of the project (Fig. 8.1). Consequently, a final demonstrator of the LEAP platform for the application in the field of manufacturing quality control has been developed. The demonstrator consisting of the rich hardware for the process surveillance, such as thermo line scanners, pertinent LEAP software components.

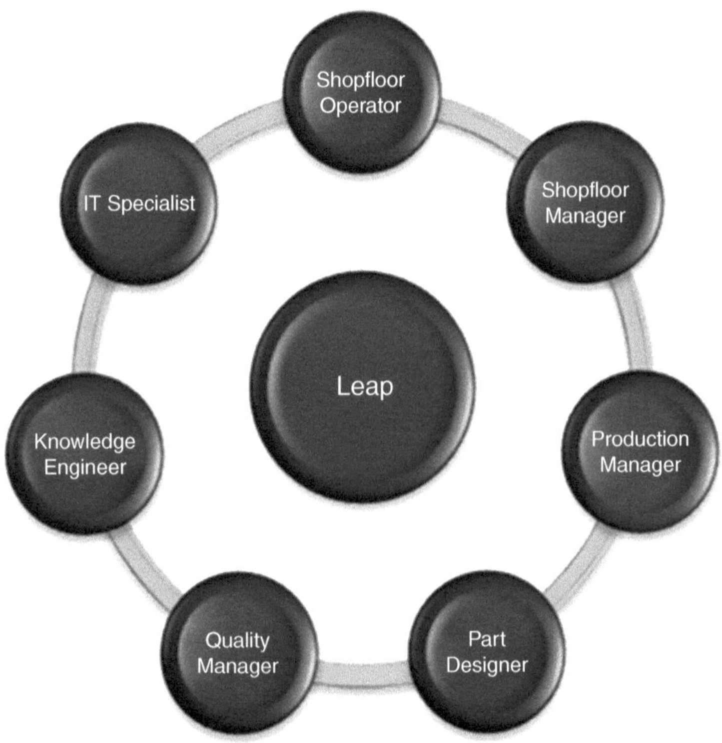

Figure 8.1 *Specific user groups who require specific data preprocessing.*

A web-based front end has been evaluated to show the potential of the developed solution and discover any open issues that have to be faced prior to the deployment of the results in the market.

8.2.1 Business Needs

Volkswagen is among the three biggest car producing companies in the world. The quality of the manufactured parts and thus of the produced cars is the key factor for being competitive to other companies. Thus, collecting data from the production lines and closing the quality control loop between production and design phases is important to improve the internal quality level. Also, the exploitation of production and quality information in departments such as financial or general management is highly valuable. The objective of this use case is to gather product and production related information along the entire production chain and use measurement parameter correlation from LEAP to reach beyond the mere extraction of scrap parts towards the identification of cause and effect in order to optimize process reliability and final quality.

As a demonstration process, we use the hot forming process which is used to produce high strength car body parts such as B-pillars. The work pieces are processed within an oven, pressed, and finally analyzed by a sensor laser scanner (Fig. 8.2). The demonstrator collects data from a production line. For each step of the production, the system collects data from sensors (eg, temperatures, pressure, forces, etc.) and links information to each produced part. Finally, to improve the quality control, the laser scanner from Trimek evaluates the compliance of a manufactured part identifying the rejected products. Analysis of the data from the field is used to improve quality control and to identify correlations between rejected parts and process parameters.

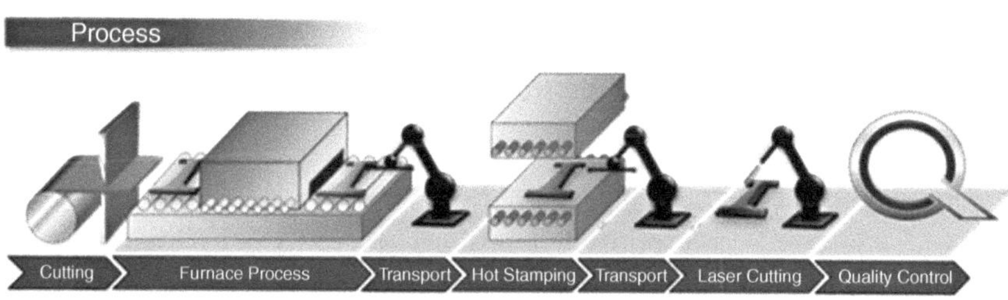

Figure 8.2 *Volkswagen demonstrator process.*

8.2.2 LEAP Application in Metrology-Driven Manufacturing

Due to the thermal surveillance of every produced part, as well as the possible geometrical probing of every third part by the Trimek system, the in-line failure detection rate was significantly increased and reached the initially defined testing ratio of 25% of the produced parts. In addition, since the related production parameters are stored, the historical data of production parameters is available for the first time and thus useful cause and effect correlation is possible, which is in stark contrast to the state prior to the project. By means of the developed concepts of the LEAP platform in the form of the web-based demonstrator, the quality control loop could be closed by the operator using the LEAP system.

Besides the direct influence on the production process, the developed platform provides a lot more beneficial futures. Due to the evaluation, the integration of the platform in addition to the current used system has the potential to speed up knowledge shearing and analysis. The daily usage of the system will simplify communication and information flow. It is very easy to get a quick overview of the relevant data, with the different developed tools and components. A short overview of potential benefits identified during the evaluation process is given as follows:

- Availability of historical process data in combination with the developed tool can be used to investigate cause and effect relations (for instance, low temperatures generate cracked products and so on) and thus contribute to the process optimization.
- Personally optimized overviews display only relevant information and will thus speed up the workflow.
- Easy access to data because it is stored in one database, which is available for every user instead of multilocation data storage.
- Improved communication between departments through the use of one system instead of the current state, where every department uses its own system.
- The production process can be monitored more efficiently from the office by means of the Real Time Monitor in terms of process parameters and by the dashboard when one is looking for the economic information on the production process.
- Overlay data on the 3D-model is very useful to recognize critical data and parts, which are displayed in red.
- Time saving in information search, since all reports and documents of components are located at Smart Search.

- Detailed view of components with name, production type, hot stamping and press parameters, scanner rep orts, and disposal causal.
- Quality improvement thanks to the Real Time Manufacturing KPIs.

8.2.3 Limitations, Barriers, and Perspectives

The evaluation of the in-use properties revealed some functionalities to be added and open issues concerning the design and usability for a final implementation. Most of the missing functionalities have been added to the demonstrator in the last project period to develop a platform that contains all desired functionalities. The following functionalities are not yet implemented and thus represent open issues in terms of basic functionalities:

- Tool that checks automatically for correlations, which is very helpful for cause and effect analysis with many different possible influence factors. The need for this functionality is eased by the incorporation of the ontology that makes suggestions for possible correlations and thus speeds up the cause–effect finding process.
- Incorporation of Microsoft Outlook into the social networking components of the LEAP platform.

Besides this, the development of a uniform design for the whole platform is necessary for a business ready solution and the implementation at possible industrial end users. In addition, explanations of color codes (legends) and possibilities for the personalization of the different components are open issues.

Since these desired changes and open issues do not require the development of new concepts, techniques or theoretical results, the research part for the desired solution (LEAP platform) is finished and the only remaining challenge is the final software development for potential end users. This is an expected result for a large scale research and integration project. It is expected that these changes can be implemented quickly.

8.3 MANUFACTURING DESIGN BASED ON PLANT LIFECYCLE COSTS

COMAU is a global supplier of industrial automation systems and services, mainly for the automotive manufacturing sector. COMAU is a proficient system integrator and it offers complete engineering solutions, from product development to manufacturing, from assistance to the production start-up phases, up to equipment and plant full maintenance activities.

The application of solutions based on LCC methodology offers a significant improvement to the production systems thanks to the possibility to select the more effective investment proposal considering a long term analysis. Life Cycle Cost (LCC) is a methodology, born in the middle 1970s in the US military, even if one its first conceptualization is in Japanese Taiichi Ohno's theorization, who is the creator of Toyota Productivity System, greatly focused on the Total Productive Maintenance (TPM) pillars.

8.3.1 Business Needs

Nowadays, despite the big steps taken by the most important end users, the LCC evaluation is made "partially" because often only some parts of the total costs are considered.

In addition, especially in companies in which the plant manager has been present for few years and whose performance is evaluated in this short period time, the approach is to give importance only, or almost, to the initial cost of investment. It is hard to demonstrate that a big initial investment can bring greater benefits in the future if the manager is no longer present in the company. Probably, for a manager assessment some criteria should be introduced considering the contributions that he or she brings in the long term to the company. If not, then there is the risk that the manager will submit negative financial statements because the long term benefits are not yet visible and the higher investment is already evident in the balance.

Again, the supplier is prompted to focus on the initial cost, both from the customer as well as competitors. For the same reason the supplier is assessed in a very short period (more or less 2 or 3 years).

So, the implementation of the LCC methodology as a way to validate the performance of manufacturing system extends the judgment horizon for both the seller and buyer, so that other significant aspects are considered by evolved companies, that are not included in the initial cost.

Moreover, the participant countries of the Kyoto agreement, which includes the European Union, are committed to ensure the respect of the maximum CO_2 emission quote attributed to each country, down to the production plants and to the companies of the single nations. Following Kyoto, the European Union delivered an important directive, which requires both new and used machines builders to look for an agreement in the short to establish actions in reducing energy consumption. Consequently, it will be not only useful but also strategically significant to evaluate the environmental impact of a technical solution for European Union, countries, regions, productive plants and manufacturing companies. In this competitive and evolved context, machine and production system builders are advantaged, if they are interested in environmental problems, encouraged by the sensitivity of end users and supported by the availability of SW systems dealing with LCC and LCA issues.

The LEAP application moved with the intent to meet these challenges and, in particular, the possibility to improve the capacity to calculate in advance the behavior of a manufacturing plant, to update continuously data banks for LCC analysis, to improve the team capacity to analyze and provide real data, and to develop some methodologies to provide anticipate information to customers.

8.3.2 LEAP Application in the Case of Plant Lifecycle Costing

The primary objective of LCC analysis is to evaluate and optimize a product's life cycle cost while satisfying specified performance, safety reliability, maintainability, and other requirements. The aim is to provide input to decision making in all phases of product's life cycle, especially in the early phases. Common types of decision, for which the LCC process can be used include, for example:

- evaluation and comparison of alternative design approaches;
- assessment of the economic viability of a project or product;
- identification of drivers and effective improvements in terms of cost;
- evaluation and comparison of alternative strategies for product usage, operation, tests, inspection, maintenance;
- evaluation and comparison of different approaches for replacement, life extension or disposal of ageing facilities;
- allocation of available funds among the competing priorities for product development;
- assessment of product assurance criteria through verification tests and its trade-off; and
- long term financial planning.

Of course, not all LCC items are identifiable by the supplier, meanwhile all items in the LCC are visible to the end user: some items are totally determined by the supplier, others from end user, and another section from both. Actually, the best approach is to radically change the approach to LCC calculation, as much as the end user and the main contractor should know the complete item set composing the whole lifecycle, otherwise it will be impossible to understand during the concept phase if one technical solution is better than another. Making this approach concrete, the end user has to give to main contractor confidential data because of privacy, politics, or administration reasons. During the concept phase, the LCC should be strategic for the future manufacturing plant, so the end user must increasingly involve the supplier to achieve this target. However, the supplier is usually only involved during the proposal phase, even if the most important choices are just taken in the previous concept phase, so the supplier must only follow the path outlined before to align the technical and economic offers. In this way, the high risk, both for end user and supplier, is to choose the solution with the cheapest initial cost but with an expensive lifecycle cost or, for evolved companies, not to take the competition because it is more expensive than others.

Nowadays, LCC and TCO studies are made during the "proposal" phase, when end users demand them. Moreover, these studies are executed only for the base solution following the initial proposal by the end user. Very rarely, a comparative study is made including different proposals and technologies in order to solve the same problem.

Surely, it is a competitive advantage to have reliable methodologies, software, and data banks that are able to execute estimations and calculations in LCC analysis during all lifecycle phases of a plant, from "concept" phase helping the end user to the "proposal" phase looking to win the competition tender, and then to the "design" phase to effectively project the LCC, and then update values and data during "build & install" and "ramp-up" and finish in "operational" phase, when it is possible to enrich and verify data banks with field data.

The main challenge consists in preparing more technical alternatives, cost estimations, and then different LCC evaluations with the same effort, in terms of time and human resources, required for the preparation of one solution as now during a proposal phase: this theme involves a lot of problems beyond the focus of LinkedDesign project and LCC methodology.

However, supposing that we have all of the data available for the LCC and LCA calculation, various "arguments" will be needed during all phases, in which LCC can be broken down for a plant, line, or machine.

For example, it is clear that the concept phase is preparatory for all phases and fundamental to determine most of the LCC calculation. In this case, it is necessary to have rough data banks because they must refer to a station, or a machine or at higher level to an entire production line as a whole. The software should be able to receive data less accurate and then execute estimations, providing an output, that should be reflected as close as with the data collected during the "operational" phase, which are the most detailed.

A satisfactory gap between the estimations during the concept phases, useful for the budget prediction, and in the operational phase is easy to answer. This is why it is fundamental to have a specific data bank (probably a SW for each LCC voice), one for concept analysis and another operational one to measure the difference.

If a machine builder, such as COMAU, should carry out a comprehensive LCC analysis, then it would be essential to have an exhaustive knowledge of the plant behavior. This is possible only if the main parameters of the plant manufacturing and organizational processes are well known, such as: organization of generic and specialized personnel; storage management, work in progress and complete product; production and logistics; fixture and metrology management; preventive and corrective maintenance organization. The variables involved in the production process for the calculation of LCC are sometimes even not clear to the customers.

In fact, some productive processes impact on both technical and management costs. The differentiation of these two entries, which is very important for LCC evaluation, is not entirely considered by the end user. This raises a risk to the machine builders during the proposal stage of the production systems that are designed without an optimized process.

In any case, each LCC solution cannot be separated to certain constraints, a machine, a line or a manufacturing plant must be designed to produce a certain number of good parts in a determined cycle time, under some limitations. Starting from this point, the study of different solutions, which differ from the initial and lifecycle costs, must achieve the productive target required by the end user. For this reason, the cheapest LCC proposal could not reach the main goal. So, the best LCC solution is to be found among the restricted set that respects the minimum number of constraints.

Nowadays, the necessary knowledge to make a good and corrective LCC analysis during the proposal phase is scrappy and incomplete.

The knowledge gaps to make a perfect and complete LCC analysis cannot be filled completely by the customer's specifics and contracts, technical offer and knowledge of own product and process standards; it is necessary a continuous intervention of experts who fill the lack or incompleteness of data, also from end user.

To make the process effective and efficient, it is necessary that any kind of information can be rapidly accessible and shareable to everyone working on LCC items as maintenance, energy consumption, and other costs. Moreover, information must be structured based on specific homogeneous data banks, which are different in form and accuracy according to the phase

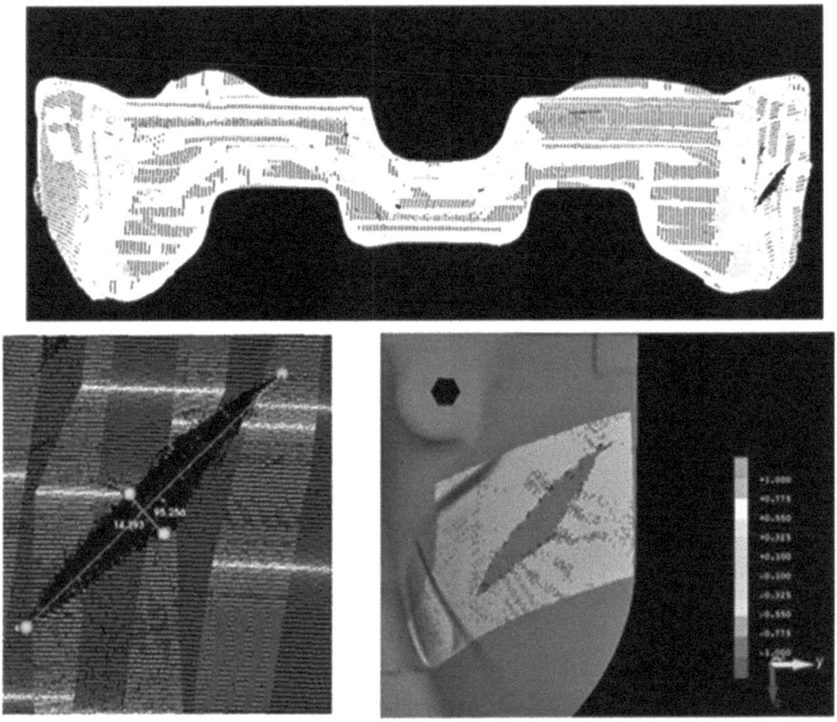

Figure 8.3 ***Metrology information generated.*** PointCloud (top), measurements (bottom left), comparison to CAD model (bottom right).

they are used for LCC analysis. The accuracy of databases will grow gradually from the concept to the design phase, for example, in the concept phase, it should be sufficient to have cost information (initial and final) for a complete manufacturing line, reasonably at station level as whole. Instead, the proposal phase must achieve details at the station group level (the set forming the station). But, in the design phase the database should refer to the components (the set forming the group), as shown in Fig. 8.3.

In conclusion, the LEAP platform aimed to support the entire process of the production system life (Fig. 8.4) in terms of decision making capability. The interest is focused, in particular, on the preliminary step that establishes the design guidelines and the product quality.

In fact, the project first addressed the following challenges for the industrial case:

- The creation and identification of the optimal lifecycle oriented solution to propose to the customer during the concept phase. Indeed, industrial systems' companies have to participate to a tender. The company

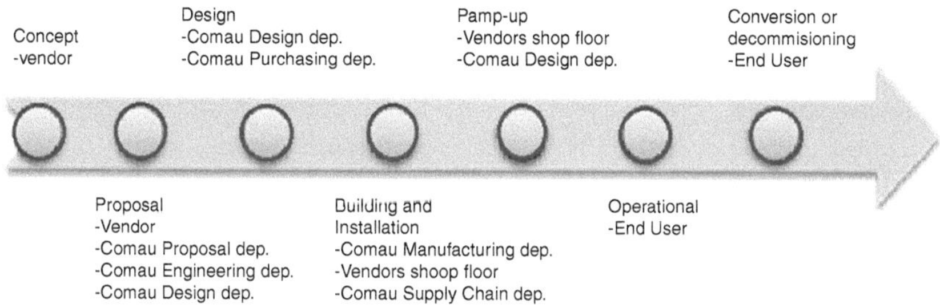

Figure 8.4 ***Product development process.***

that provides the best solution in terms of life cycle cost (and/or life cycle environmental impact), which is one of the key drivers for the choice, wins the order.

- The collection of data from the field is required to increase the knowledge of their systems. Indeed, being industrial systems as one-of-a-kind solutions, designers and system engineers have not had a complete knowledge on their systems and their behavior during the use phase.

8.3.3 Limitations, Barriers, and Perspectives

The results of the project, although of considerable significance, have been achieved in the context of a prototype level.The conditions were definitely more controlled and binding with respect to an industrial reality, much more complex and unpredictable, but also because until now the stakeholders have not addressed the issues, developed in the project, with the necessary insights and awareness. In fact, one of the challenges of the LCC/LCA Service, was the environmental impact evaluation. In the past, no one has conducted an exhaustive LCA analysis and, therefore, there were difficulties collecting data about the environmental dimension. So, a simplified Life Cycle Assessment was performed using limited data.

Moreover, for the prototype application, the LCC/LCA Service has been realized using a previous proposal so that we could evaluate whether LCC/LCA Service returned appropriate solutions. For a conclusive assessment, the tool should run in normal stress conditions, that allows us to evaluate the effective support in decision making process according to stringent timelines.

Another important tool, the PLM Data Service, was implemented in a laboratory, hence the collected data does not contain all of the events that are normally present in a real manufacturing line, the retrieve time is also short compared to the duration of an installation. In any case, this tool enabled us to understand how PLM Data Service worked and how we could integrate it with the PLC (Programmable Logic Controller) of the different stations.

These and other limitations are easily overcome if it is possible to establish a trust relationship with end users that will allows us to implement the LEAP platform in their facilities and so share operation data. Obviously, a confidentiality agreement between the stakeholders will be necessary. This is the real barrier for the exploitation of the results in the automotive industry.

A significant feature of the LEAP platform is the possibility to share on the web reporting documents, that could be extremely reserved. The confidentiality of the documents by unauthorized users must be properly ensured.

In the context of current trends in the automotive sector towards high reliability, safety, high efficiency, and low environmental impact transportation, the end user will benefit greatly from a deep collaboration with production system builders. There is, therefore, a need for an ICT structure support in the area of product development that allows us to coordinate the activities of the process and to reduce or even eliminate downtime for research or awaiting information. A platform to develop the factory and the process as for create the final product, from the first conceptual layout until the finish product. In this way, it is clear the configuration of the so-called "Virtual Factory," based on research, learning and integrated into the physical factory, which develops new products over the time, develops processes, test products and processes in the production factory, and plans and develops process variants. The future developments of the LEAP platform should be part of this innovative path.

8.4 KNOWLEDGE REUSE IN AUTOMATED DESIGN FOR OFFSHORE ENGINEERING

Aker Solutions is a global provider of products, systems, and services to the oil and gas industry. Its engineering, design and technology bring discoveries into production and maximize recovery. The company operates in over 20 countries and it is typical for larger projects to include two to four different geographical locations and employ thousands of engineers at peak manning.

Aker Solutions sees great potential for improvement through operational excellence, standardization, new technology, and innovation. In the

LinkedDesign project, the two main topics for Aker Solutions have been Knowledge Based Engineering (KBE) and "Virtual Office" (in ref. to Chapter 7), effective and efficient collaborative engineering across different locations.

8.4.1 Business Needs

Using KBE solutions saves time and cost by allowing automated creation of design models based on governing parameters and rules. To arrive at a KBE solution, a suitable project must be initiated based on identified and justified scope and business value. At first, the problem has to be analyzed to gain deeper understanding of required inputs, outputs and process steps. Based on this preparatory work, software is designed and implemented and finally introduced to engineers and designers that utilize the developed tools to increase efficiency in project execution. Training, support, and follow-up are key activities at this subsequent stage, which is also where the business value can be realized based on investments from the earlier stages. Experiences in Aker Solutions indicate several challenges to scaling up KBE to a global and multidiscipline engineering organization. A summary of identified challenges brought to the LinkedDesign consortium is listed in the following:

- *Increase the transparency of knowledge embedded in KBE tools.* In a typical KBE tool, engineers can see an initial state of a product model, manipulate a set of input parameters, and observe the resulting changes in the model. But the process by which those changes were arrived at is not available to them since the engineering and domain related knowledge is encapsulated in software code (black box). In some cases documentation exists but since the design rules vary with time and context (such as client and project specific requirements), it can be challenging to keep documentation relevant across all this variation. In cases where correct and relevant documentation do not exist, the users are left with the option of deducing the process based on experimentation with inputs and outputs or requesting explanations from KBE developers.
- *Traceability.* Even in cases where the design process is transparent, it is still desirable to trace the KBE code back to a rationale for the implemented rule. This increases the engineers' trust in the tool and process, and reduces the effort required to validate rules for a different context or if the results of applying a set of rules are challenged.
- *Reduce complexity of knowledge capture and reuse.* Current practices for knowledge capture are not standardized and the process and outcomes are highly dependent on the individuals involved. It is hard to use

standardized formats for knowledge capture because the structure and formal requirements of such formats may get in the way of the knowledge transfer from "teacher" to "learner." On the other hand, knowledge reuse through a KBE system requires that the knowledge is formalized.

- *Tool integration.* The barrier to start using and benefiting from KBE tools is significantly lower, if the functionality can be provided in the tools that are already being used, rather than introducing a new tool that requires training and may use different conventions than the familiar tools (eg, default coordinate-systems, UI elements, etc.)
- *Collaboration in large virtual engineering teams.* Aker Solutions engineering projects are typically coordination- and collaboration intensive multidiscipline projects that involve several international locations. Combining collaboration enabling technologies and new work and management practices to support knowledge sharing in effective and efficient multidisciplinary work sessions will be a competitive advantage for Aker Solutions.

8.4.2 LEAP Application in Knowledge Reuse

The Knowledge Brief (K-Brief) is the basic format for knowledge capture and reuse in LEAP. The concept is gleaned from Lean Product Development, and its roots can be traced to the Toyota Production System and A3 reports.

A few alternatives were investigated before settling on K-Briefs as the basic format for knowledge capture and reuse in LEAP. Initial attempts with more specialized approaches have indicated that too much effort would be spent on adapting formats and routines to specific applications and disciplines, while the Knowledge Brief is a general tool for knowledge capture and reuse, which is endorsed by a broad range of practitioners.

Retrieval of relevant K-Briefs according to context is supported by adding metadata such as the product, discipline, project, relevant execution stages, and so on to K-Briefs. However, this requires a lot of metadata to be manually filled in before a K-Brief is accepted, which reduces flexibility and simplicity and is a barrier to use.

The minimal set of metadata that must be included is *author*, *theme*, and *date of last update*. These data should be visible in the document itself, rather than in the document store only (eg, allowing comparison with a printed version of the document). Additional metadata can be added according to the particular purpose but should be limited to avoid unnecessary increase in user effort. The impact of missing or wrong input must also be considered.

In LEAP, an add-in was developed that allows authoring K-Briefs in OneNote, which are published to central storage by a single click. The add-in is based on a particular template for K-Briefs and automatically extracts data from the document, and as part of the publishing process would upload metadata to the Smart Link Store (in ref. to Chapter 2). When the K-Brief is published to the Smart Link Store, a field in the template is filled in with links to related K-Briefs that also have been published to the store.

Further extensions to the K-Brief concept based on leveraging metadata have been proposed, for example, "Virtual K-Briefs" that could assemble relevant smaller pieces of knowledge in a similar format based on metadata and a provided context. In other words, the content is merged from different sources into the K-Brief format and displayed.

The K-Brief has been introduced in Aker Solutions and is being used as part of our current approach to KBE, mainly as a documentation format. The Virtual Obeya prototype and concept is suitable to address many relevant collaboration scenarios in Aker Solutions, where simultaneous access to comprehensive set of data supports information transfer and decision-making processes across geographical and cultural borders.

In LinkedDesign, an add-in was created for Aveva PDMS, which is a popular a CAD system that is used in the design and engineering of process plants that allowed documentation generated from RIF/SysML to be retrieved from the PDMS product hierarchy.

In our application scenario, the RIF model was used to establish a connection between representations of classes and objects in a KBE system, on one side, and documentation in the form of K-Briefs, on the other side.

A model was exported from a KBE system into a CAD system while maintaining a mapping to its corresponding RIF/SysML model, with relevant K-Briefs linked to model elements. The K-Briefs included further references to documents and people. In this way documentation for rules, project settings and other aspects impacting the configuration of the model can be accessed through RIF directly from the KBE model.

8.4.3 Limitations, Barriers, and Perspectives

During prototyping in LinkedDesign, a modeling-tool called TopCased was mainly used to create, read, manipulate, and export the models. An important reason for this was the fact that TopCased is free and open-source. Some experimentation with other tools, suggests that commercial tools exist that are better suited for the task based on user friendliness and more comprehensive feature sets.

Since RIF (in ref. to Chapter 6) is based on SysML, the interchange of models between tools is in principle supported by the XMI standard. However, in practice the models created in the project could not be interchanged between tools. Unfortunately, it seems that this is a generally known problem with XMI. RIF is currently being developed further with the eventual goal of achieving a standard for the description and exchange of KBE models.

The Virtual Obeya prototype was not integrated with the software systems that are generally used in Aker Solutions (MS Office, MS SharePoint, various engineering systems). This limited the ability to evaluate Virtual Obeya in more production-like settings. Some evaluation feedback also points to a need for better differentiation from screen sharing and SharePoint in terms of unique selling points.

8.5 CONCLUSIONS

In this chapter, three use cases are described in details, giving the AS-IS company's processes as they were prior to LinkedDesign project, as well as the issues the LinkedDesign was meant to address. It was shown that these required improvements were successfully conducted and that wide diversity of challenges can be addressed through generic modules of LEAP platform. Data federation and knowledge management contributed to metrology systems in automotive industry by enabling better access to information as well as data analytics tools. The same principles were used to improve decision support mechanisms in the plant design stage, when all of the decisions have a strong effect on the future lifecycle stages of a product. Finally, knowledge management was formalized through the concept of Knowledge Briefs that brings automated and simplified structuring and exchange of vital product information.

INDEX

H

I

J

K

L

X

FSC
www.fsc.org
MIX
Papier | Fördert gute Waldnutzung
FSC® C083411

Kolibri 360 GmbH
Ferdinand-Jühlke-Straße 7
99095 Erfurt, Deutschland
produktsicherheit@kolibri360.de